Instructor's Manual
to accompany

PRINCIPLES OF
MODERN
CHEMISTRY

**FOURTH
EDITION**

OXTOBY / GILLIS / NACHTRIEB

DAVID W. OXTOBY
The James Franck Institute,
University of Chicago

SAUNDERS GOLDEN SUNBURST SERIES
Saunders College Publishing
Harcourt Brace College Publishers

Fort Worth Philadelphia San Diego New York Orlando Austin
San Antonio Toronto Montreal London Sydney Tokyo

Printed in the United States of America

ISBN 0-03-024752-7

901 023 7654321

CONTENTS

TO THE INSTRUCTOR

There are many ways to approach the teaching of chemistry, and it is probably correct to say that there is no one best way. How an instructor of chemistry organizes the subject reflects, to a considerable degree, his or her view of how it relates to the other sciences. We see it as the central science that furnishes other sciences with concepts and tools of investigation, and that receives from them in return the stimulus of new applications and new phenomena.

This fourth edition of PRINCIPLES OF MODERN CHEMISTRY follows the overall organization of the first three editions, and represents our approach to the teaching of a first course in chemistry at the college level. We are aware that it lays open our prejudices and biases in this endeavor. One of these is that the subject is enriched for the student when it is presented in roughly chronological order. That is to say, we think it makes sense to start in Chapter 1 with the atomic theory of matter as it developed with Lavoisier and Dalton, rather than with the electronic structure of the atom. This helps to give the student a sense of how science develops . . . how it learns from its mistakes and misconceptions, how the give-and-take of experiment and theory are vital to its progress. By emphasizing macroscopic concepts before microscopic ones, we hope to show the student that chemistry is a real, tangible subject, not just a theoretical construct.

Another of our convictions is that a choice need not and should not be made between "descriptive chemistry" and "physical chemistry," a fruitless debate that has divided teachers of chemistry for forty years. There is no conflict between these aspects of chemistry, but rather only an opportunity to give meaning to the facts of laboratory observation by appeal to theory on the one hand, and to illustrate and provide examples for the testing of theory on the other.

Of the students we had in mind when we wrote PRINCIPLES OF MODERN CHEMISTRY, perhaps only 10 to 20% will become chemists. Many of the others will pursue admission to the medical profession, and others will become economists, mathematicians, physicists, concentrators in English literature or other fields. In another college the fraction of students in the beginning chemistry course might be weighted more toward engineering. We believe these to be minor considerations because there are often changes of concentration by the student, and more fundamentally because there is ideally one kind of chemistry course for any serious student, irrespective of the profession he or she hopes to enter.

Having stated our prejudices about the "proper" way to teach chemistry, we should point out that we have tried not to be too dogmatic in our presentation of the subject, and we have written the book in such a way that the instructor should have considerable flexibility in choice of subjects, course organization, and level of the material. Below we outline several alternatives to our method of organization that may be suitable for different courses.

1. A course with "descriptive chemistry" emphasis

Such a course would be selective in its coverage of the first nineteen chapters, and would devote additional time to the last nine chapters. Any or all of the sections entitled "A Deeper Look" could be omitted in such a course without loss of continuity. Other sections that could be left out include 4.7, 6.4 to 6.6, 7.6, 10.7, 11.2, 11.6, 11.8, Chapter 14, Chapter 16, 18.4, 18.5, 18.6, 19.5, and 19.6.

2. A course with "physical chemistry" emphasis

This course might cover the material in chapters 1 through 16, 18, and 19 rather fully, including most of the "Deeper Look" sections. Only selected material from the last six chapters might be included. Chapters 20-22, for example, outline the chemistry of some important groups of elements, and one might serve as a representative application of chemistry to "real world" problems. Chapters 23 and 24 extend the study of simple solids to important naturally ocurring and synthetic solid materials.

3. A one semester course in chemistry for engineers

We find it difficult to imagine covering only half the material in a general chemistry course, but we recognize that some degree programs do not permit more time than this. Our recommendation for an intensive one semester course for engineers would be to cover most of the first thirteen chapters in this book, together perhaps with Chapter 19, unless that material is taught elsewhere in the curriculum. This course would include significant emphasis on macroscopic aspects of chemistry, together with an introduction to chemical periodicity and simple models of chemical bonding in Chapter 3.

4. A course with early emphasis on atomic structure and bonding

We have stated our prejudice above for emphasizing macroscopic and experimental aspects of chemistry before presenting the quantum theoretical foundations of the subject. Nonetheless, we recognize that some instructors will prefer to discuss atomic structure and bonding more extensively early in the course. To us, this material is in fact technically easier, although conceptually more difficult, than some of the material from the first thirteen chapters of the book. We have written the book in such a way that a body of material on atomic structure and chemical bonding (Chapters 15 and 16) may be presented earlier in the course. For example, these chapters could be inserted after Chapter 3 or after Chapter 6.

SUGGESTED LECTURE DEMONSTRATIONS

Demonstration experiments are an important part of the students' learning process in introductory college chemistry (and of the instructor's teaching process as well). They remind both that chemistry is an experimental activity at the outset. They also furnish a sense of reality and relevance to happenings in the world around us. Without them, the subject matter is often seen by the student as "theoretical," unrelated to the processes that go on all around, and consequently uninteresting.

Demonstration experiments should be relatively short and to the point. They should relate to the topic about to be considered in the lecture or discussion. Their purpose is not so much to *answer* questions as to *raise* them and to arouse interest in students' minds. Some demonstrations are dramatic and impressive, but others that are lower key can be equally successful in challenging students to think.

The following are suggestions for experiments that have proved to be useful. References are made to articles in the Journal of Chemical Education, to Tested Demonstrations in Chemistry by Hubert N. Alyea and Frederic B. Dutton (published by the Division of Chemical Education of the Americal Chemical Society), and to the four volumes of Chemical Demonstrations by Bassam Z. Shakhashiri (published by the University of Wisconsin Press). An instructor is invited to select demonstrations from the following list, or to choose others from the hundreds described in the above references.

Chapter 1
Mole-sized samples. Display one-mole quantities of elements (iron, sulfur, lead, etc.) and compounds (sucrose, salt). Use a cardboard cube representing 22.4 L of an ideal gas at STP, or place 44 g of dry ice into a plastic garbage bag, allowing it to sublime and inflate the bag.

Law of combining volumes. The volume of SO_2 after combustion of sulfur in oxygen equals the original volume of oxygen [Shakhashiri, Vol 2, p. 190]

Density of liquids. Measure the relative densities of oil and water using a U-tube. [Test. Dem., p. 67]

Reaction of sodium with water. Use a Petri dish on an overhead projector, with a trace of phenolphthalein indicator put into the water before the sodium.

Reaction of calcium hydride with water. Use a Petri dish on an overhead projector.

Chapter 2
Perform a reaction. Examples include the reaction of sodium and chlorine [Shakhashiri, Vol 1, p. 61], of a copper penny with nitric acid [Shakhashiri, Vol 1, p. xiv], or of aluminum with bromine [Shakhashiri, Vol 1, p. 68]. There are many other possibilities.

Physical and chemical change. Ignite a length of magnesium ribbon [Shakhashiri, Vol 1, p. 38], observing physical and chemical changes (melting of the ribbon, combustion, deposition of MgO on forceps, mixing of heated air and smoke into the room).

Combustion of propane. This demonstrates the importance of correct stoichiometry. [J. Chem. Ed. **64**, 894]

Chapter 3

VSEPR geometries. Use sets of balloons of identical size and shape to illustrate geometries. Join mounts with tape.

Oxidation states of vanadium. [Test. Dem., p. 171]

Ionic versus covalent compounds. Compare the qualitative differences in conductivity of plastic, solid NaCl, solid $NaCH_3COO$, wax, distilled water, and $NaCl(aq)$ using a conductivity apparatus in which a large light bulb glows to indicate passage of a current. The $NaCH_3COO$ and wax can subsequently be melted in a crucible and tested.

Chapter 4

Mercury barometer. Show that the height of the mercury does not depend on the shape of the tube. [Shakhashiri, Vol. 2, p. 9]

Boyle's law. [Shakhashiri, Vol. 2, p. 14]

Thermal expansions of gases. Shrivel balloons with liquid nitrogen. [Test. Dem., p. 220]

Graham's law of effusion. Times measured for effusion of gases into a vacuum. [Shakhashiri, Vol. 2, p. 72]

Diffusion in gases. Diffusion of $HCl(g)$ and $NH_3(g)$ in a glass tube. [Shakhashiri, Vol. 2, p. 59]

Many other demonstrations are given in Shakhashiri, Vol. 2.

Chapter 5

Five liquid phases. [J. Chem. Educ. **64**, 694 (1987)]

Undercooling of a liquid. Use thymol. [Shakhashiri, Vol. 1, p. 34]

Vapor pressure. Show variation with temperature. [Shakhashiri, Vol. 3, p. 252]

Cold boil. Put 10 mL of warm water in a 50 mL plastic syringe, seal the end and reduce the pressure inside the syringe by pulling the plunger nearly out. The water boils. See also Shakhashiri, Vol. 2, p. 81.

Chapter 6

Titrations. Perform a portion of the titration of Fe^{2+} with MnO_4^-. Bright lighting from the side or bottom shows the color changes well.

Dissolution of gases. Show the dissolution of gases in liquids by doing the ammonia fountain demonstration. [Shakhashiri, Vol. 2, p. 205]

Raoult's law. [Test. Dem., p. 195]

Distillation. [Shakhashiri, Vol. 3, p. 258; Test. Dem., p. 11]

Tyndall effect. Light is scattered by colloidal particles. [Shakhashiri, Vol. 3, p. 353; Test. Dem., p. 169; J. Chem. Educ. **65**, 623 (1988)]

Many other demonstrations are given in Shakhashiri, Vol. 3, Section 9.

Chapter 7

Thermite reaction. Observe cautions in Shakhashiri. [Shakhashiri, Vol. 1, p. 271]

Heat of neutralization. Measure temperature changes upon diluting or neutralizing acidic or basic solutions. [Shakhashiri, Vol. 1, p. 15]

Endothermic dissolution reaction. Dissolve ammonium nitrate in water. [Shakhashiri, Vol. 1, p. 8; J. Chem. Educ. **65**, 267 (1988)]

Endothermic acid-base reaction. Reaction of $Ba(OH)_2 \cdot 8\,H_2O$ with NH_4NO_3. [Shakhashiri, Vol. 1, p. 10]

Rubber band Carnot cycle. Note temperature changes when a rubber band is rapidly stretched or released. [Test. Dem., p. 200; J. Chem. Educ. **29**, 405 (1952)]

Chapter 8

Spontaneous endothermic reactions. [J. Chem. Educ. **46**, A55 (1969)].

Irreversible gas expansion. Use NO_2 gas. [Test. Dem., p. 160]

Spontaneous gas dissolution. Use HCl or NH_3, showing the fountain effect. Work to lift water is provided by the free energy change. [Shakhashiri, Vol. 2, p. 205]

Chapter 9

NO_2-N_2O_4 equilibrium. Use sealed tubes at several temperatures with an overhead projector. [Shakhashiri, Vol. 2, p. 180; Test. Dem., p. 19, 131, 167]

Le Chatelier's principle. [Test. Dem., p. 19, p. 221; J. Chem. Educ. **47**, A735 (1970)]

Chapter 10

Indicator colors. Prepare and mix colored solutions of indicators, using an overhead projector. [Shakhashiri, Vol. 3, p. 33]

Natural indicators. Grape juice works well, as does red cabbage extract. [Shakhashiri, Vol. 3, p. 50; Test. Dem., p. 204]

CO_2 production. React an acid with a carbonate or bicarbonate. [Shakhashiri, Vol. 3, p. 96; J. Chem. Educ. **62**, 1108 (1985)]

Weak acids. Use conductivity to compare weak and strong acids. [Shakhashiri, Vol. 3, p. 140]

Hydrolysis. Demonstrate acid-base properties of salts. [Shakhashiri, Vol. 3, p. 103; Test. Dem., p. 62]

Buffer action. [Shakhashiri, Vol. 3, p. 173; Test. Dem., p. 62, 128, 155, 210; J. Chem. Educ. **60**, 493 (1983), **62**, 608 (1985)]

Neutralization of an antacid. Use a buret or a graduated cylinder. [Shakhashiri, Vol. 3, p. 162]

Acidic and basic properties of oxides. [Shakhashiri, Vol. 3, p. 109].

Properties of acids and bases. [Shakhashiri, Vol. 3, p. 70]

Lewis acids and bases. Demonstrate the reaction of $SO_2(g)$ or $CO_2(g)$ with $CaO(s)$. [Test. Dem., p. 147]

Many other demonstrations are given in Shakhashiri, Vol. 3, Section 8.

Chapter 11

Chromatography. [J. Chem. Educ. **59**, 1042 (1982); **62**, 530 (1985)].

Precipitation reactions. Use medicine droppers to position separate small puddles of reactive solutions such as $Pb(NO_3)_2(aq)$ and $K_2CrO_4(aq)$ on a Petri dish on an overhead projector. Join the two solutions into a third separate puddle of water by gently dragging the tip of a dropper from one to another. As the solutions mix, the precipitate (of $PbCrO_4$ in this case) appears as a haze and thickens with time in various interesting patterns.

Crystals of lead iodide. [Test. Dem., p. 45]

Solubility product constant. Estimate K_{sp} for $Cd(OH)_2$. [Test. Dem., p. 173]

Common ion effect. [Test. Dem., p. 19, 86]

AgCl-AgCrO$_4$ equilibrium. [J. Chem. Educ. **54**, 618 (1977)]

Orange tornado. Show effects of complex ion formation on solubility of mercury(II) salts. [Shakhashiri, Vol. 1, p. 271]

Ni(II) complexes and precipitates. [Shakhashiri, Vol. 1, p. 299]

Amphoteric metal hydroxides. Show role of complex ions in solubility of metal hydroxides. [Shakhashiri, Vol. 3, p. 128]

Precipitation of arsenic(III) Sulfide. A clock reaction [Shakhashiri, Vol. 4, p. 80]

Many other demonstrations are given in Shakhashiri, Vol. 1, Section 4.

Chapter 12

Redox reaction. The simplest involve the direct combination of the elements. [Shakhashiri, Vol. 1, pp. 53, 55, 64, 66, 68, 121]

Redox reaction: the beating heart. Oxidation of mercury. [Test. Dem., p. 143]

Generate electric current. Use $Mg/Mg^{2+}//Cu^{2+}/Cu$ cell to ring a door bell or set off a photoflash bulb. [Test. Dem., p. 17, 150]

Galvanic cells using vegetables. [J. Chem. Educ. **65**, 727 (1988)]

Electrolysis of water. [J. Chem. Educ. **63**, 809 (1986)]

Fuel cell. [J. Chem. Educ. **65**, 725 (1988)]

Corrosion. [Test. Dem., p. 104, 171; J. Chem. Educ. **58**, 505, 802 (1981); **62**, 531 (1985); **65**, 156 (1988)]

For many useful electrochemistry demonstrations, see Shakhashiri, Vol. 4, Section 11.

Chapter 13

Carbon dioxide reaction rates. Rates of acid-base reactions in aqueous solution for the catalyzed and uncatalyzed reaction of CO_2 with water to make H_2CO_3. [Shakhashiri, Vol. 2, p. 122]

Effect of concentration on rate. [Test. Dem., p. 84; J. Chem. Educ. **42**, A607 (1965); **62**, 153 (1985)].

Effect of temperature on rate. [J. Chem. Educ. **58**, 384 (1981)]

Catalytic oxidation of $MnSO_4$. [Test. Dem., p. 206]

Briggs-Rauscher reaction. Oscillating reaction. [Shakhashiri, Vol. 2, p. 248]

Other demonstrations of oscillating reactions are given in Shakhashiri, Vol. 2, Section 7.

Clock Reactions. See Shakhashiri, Vol. 4, Section 10.

Chapter 14

Gas Discharge Tube. Show how a gas glows under a high voltage when its pressure is reduced sufficiently. [Shakhashiri, Vol. 2, p. 90]

Gamma radiation. Use a Geiger counter and a meter stick to demonstrate the inverse square law for gamma radiation. [Test. Dem., p. 22].

Beta particles. [Test. Dem., p. 21]

Cloud chamber. [Test. Dem., p. 21, 90]

Chapter 15

Flame tests. Try Na, Ca, Sr, Li, Cu, Ba. [Test. Dem., p. 22, 91; J. Chem. Educ. **65**, 452 (1988)]

Spectroscopy with large grating. [J. Chem. Educ. **65**, 266 (1988)]

Chapter 16

Paramagnetism of liquid oxygen. Pour liquid oxygen between the poles of a magnet. [Shakhashiri, Vol. 2, p. 147; J. Chem. Educ. **57**, 373 (1980)]

Singlet molecular oxygen. Use the reaction of hydrogen peroxide and hypochlorite ion; red light is emitted in collision of pairs of these excited-state molecules. [Shakhashiri, Vol. 1, p. 133]

Chemiluminescence. [Shakhashiri, Vol. 1, p. 125]

Photochemical reaction. Use a projector to initiate the reaction of hydrogen and chlorine. [Shakhashiri, Vol. 1, p. 121]

Absorption of UV light by ozone. [J. Chem. Educ. **66**, 338 (1989)]

Chapter 17

Cracking of mineral oil. Detect formation of alkenes by their addition reactions with bromine. [Test. Dem., p. 184]

Hydrolysis of 2-chloro-2-methylpropane. A clock reaction. [Shakhashiri, Vol. 4, p. 56]

Aldehyde-acetone condensation. A clock reaction. [Shakhashiri, Vol. 4, p. 65]

Chapter 18

Formation of a coordination complex. Make $[Cu(NH_3)_4]^{2+}$, and display on an overhead projector. [Test. Dem., p. 22]

Complexes of cobalt(II). [Shakhashiri, Vol. 1, p. 280]

Complexes of nickel(II). [Shakhashiri, Vol. 1, p. 299; J. Chem. Educ. 57, 900 (1980)

Chiral geometry. Use models to demonstrate.

EDTA complexes. [Test. Dem., p. 132]

Paramagnetic substances. Prepare samples of iron (ferromagnetic), various iron(II) and iron(III) compounds (paramagnetic), and water, salt, and so forth (diamagnetic) in large test tubes suspended from a stand. Test with a powerful horseshoe magnet.

Many other demonstrations are given in Shakhashiri, Vol. 1, Section 4.

Chapter 19

Unit cell models. Construct models from balsa wood of units cells in the seven crystal systems. [J. Chem. Educ. **66**, 73 (1989)]

Crystal growth. Use gels to grow crystals. [Shakhashiri, Vol. 3, p. 372]

Thermal expansion of liquids. Compare relative values for several liquids. [Shakhashiri, Vol. 3, p. 234]

Surface tension of liquids. Exhibit properties of oils or detergents on water. [Shakhashiri, Vol. 3, p. 301]

Hydrogen bonding in liquids. [Test. Dem., p. 132, 143]

Ice bomb. Demonstrate expansion of water on freezing. [Shakhashiri, Vol. 3, p. 310]

Color centers in KCl. [Test. Dem., p. 132]

Chapter 20

Preparation and properties of hydrogen. React zinc with acid. [Shakhashiri, Vol. 2, p. 128]. See also Vol. 4, p. 130.

Separation of ore by flotation. [Test. Dem., p. 25]

Reduction of an ore. [Test. Dem., p. 25]

Electrolysis of metal salts. [Shakhashiri, Vol. 4, p. 205]

Electroplating of Cu, Ni, Cr, Ag, and Zn. [Shakhashiri, Vol. 4, pp. 212-246]

Chapter 21

Preparation of NO. [Shakhashiri, Vol. 2, p. 163]

Preparation of ammonia. [Shakhashiri, Vol. 2, p. 202]

Oxidation of ammonia. Use a hot platinum wire to oxidize ammonia to NO and NO_2, as in the Ostwald process. [Shakhashiri, Vol. 2, p. 214]

Sulfur and polysulfides. React sulfur with concentrated NaOH. [Shakhashiri, Vol. 3, p. 58]

Mineral acids. Compare reactions of sulfuric, nitric, and phosphoric acids (together with hydrochloric). [Shakhashiri, Vol. 3, p. 70]

Heat of dilution of sulfuric acid. Illustrates care that must be taken in adding acid to water. [Shakhashiri, Vol. 1, p. 17]

Combustion of guncotton. [Shakhashiri, Vol. 1, p. 43]

Combustion of white phosphorus. [Shakhashiri, Vol. 1, p. 74]

Decomposition of nitrogen triiodide. [Shakhashiri, Vol. 1, p. 96]

Disproportionation of sodium thiosulfate. A clock reaction. [Shakhashiri, Vol. 4, p. 77]

Chapter 22

Preparation of HCl. Use the traditional reaction of sulfuric acid with sodium chloride. [Shakhashiri, Vol. 2, p. 198]

Preparation of chlorine. Use the reaction of MnO_2 and hydrochloric acid. [Shakhashiri, Vol. 2, p. 220]

Reactions of chlorine with metals. [Shakhashiri, Vol. 1, p. 61, 64, 66]

Etching glass. Generate hydrogen fluoride and use it to etch a watch glass. [Shakhashiri, Vol. 3, p. 80]

Oxidation of iodide with hydrogen peroxide. One of several clock reactions involving iodide. [Shakhashiri, Vol. 4, p. 37]

Chapter 23

Silicate garden. Grow colored silicates. [Test. Dem., p. 33]

Superconducting ceramic. Levitation of a magnet on a disk of $YBa_2Cu_3O_{9-x}$ at liquid nitrogen temperature. Kit may be purchased from Institute for Chemical Education (Project 1-2-3 Levitation Kit), University of Wisconsin-Madison.

Chapter 24

Phosphors. Prepare several luminescent compounds. [Test. Dem., p. 212]

Chapter 25

Dehydration of sugar. Use concentrated sulfuric acid to leave a black solid. [Shakhashiri, Vol. 1, p. 77]

Burning polyvinyl chloride. Illustrate the hazards of direct incineration. [Shakhashiri, Vol. 3, p. 116]

Physical properties of polymer solutions. Polymer solutions ofter have unexpected flow properties compared to simple liquids. [Shakhashiri, Vol. 3, p. 333, 335]

Slime. Gelation of polyvinyl alcohol. [Shakhashiri, Vol. 3, p. 362]

Nylon rope trick. Pull a thread of nylon polymer from the interface between two liquids. [Shakhashiri, Vol. 1, p. 213]

Cuprammonium rayon. Rayon threads are formed by adding a solution containing copper(II) ions, ammonia, and paper to an acid bath. [Shakhashiri, Vol. 1, p. 247]

Many other demonstrations are given in Shakhashiri, Vol. 1, Section 3.

Chapter 1

The Nature and Conceptual Basis of Modern Chemistry

1-2 "Absolute" means pure, so absolute alcohol is a substance; milk is a mixture; copper wire is a substance. Rust is a mixture (the reason for this last answer is discussed on p. 6 of the text in the context of salt versus sodium chloride). Barium bromide is a substance. Concrete, baking soda, and baking powder are all mixtures.

Absolute alcohol and barium bromide are compounds; copper wire is an element. All of the mixtures are heterogeneous.

1-4 Proving that a material is <u>not</u> an element involves finding one of the many possible ways of breaking it down into simpler substances. To prove that a material <u>is</u> an element requires showing that there is no way to decompose it. It is always possible that trying a new method will convert it into simpler substances.

1-6 The ratio of the mass of tellurium to the mass of hafnium in this compound is

$$\frac{\text{mass Te}}{\text{mass Hf}} = \frac{31.5 \text{ g Te}}{25.0 \text{ g Hf}} = 1.26 \frac{\text{g Te}}{\text{g Hf}}$$

Because the compound from the rock is identical, it contains Te and Hf in the same ratio.

$$\text{mass Te} = 0.125 \text{ g Hf} \times \left(1.26 \frac{\text{g Te}}{\text{g Hf}}\right) = 0.158 \text{ g Te}$$

The compound may of course contain any number of other elements.

1-8 (a) In each case the mass of fluorine that combines with 1.0000 g of iodine is simply the mass percentage of fluorine divided by the mass percentage of iodine. This is best proved by considering samples of the compounds that have masses of 100.000 g. The masses contributed by each element in the compounds are then very easily computed. The ratios in the last column of the following table, which are formed by the indicated divisions, are the answers.

$$\text{Compound 1} \quad 13.021 \text{ g F}/86.979 \text{ g I} \quad 0.14970 \text{ g F}/ \text{ g I}$$
$$\text{Compound 2} \quad 30.993 \text{ g F}/69.007 \text{ g I} \quad 0.44913 \text{ g F}/ \text{ g I}$$
$$\text{Compound 3} \quad 42.809 \text{ g F}/57.191 \text{ g I} \quad 0.74853 \text{ g F}/ \text{ g I}$$
$$\text{Compound 4} \quad 51.171 \text{ g F}/48.829 \text{ g I} \quad 1.04796 \text{ g F}/ \text{ g I}$$

(b) The law of multiple proportions involves the ratio of these ratios. Simply divide all four of the answers in part (a) by the smallest of the answers. The results are: 1.0000 for compound 1; 3.0002 for compound 2; 5.0002 for compound 3; 7.0004 for compound 4. These equal the small whole numbers 1, 3, 5, and 7 within the precision of the data.

1-10 As in problem 1-8, we calculate the masses of chlorine per gram of tungsten in the four compounds. These are

$$\text{Compound 1} \quad 27.83 \text{ g Cl}/72.17 \text{ g W} \quad 0.3856 \text{ g Cl}/ \text{ g W}$$
$$\text{Compound 2} \quad 43.55 \text{ g Cl}/56.45 \text{ g W} \quad 0.7715 \text{ g Cl}/ \text{ g W}$$
$$\text{Compound 3} \quad 49.09 \text{ g Cl}/50.91 \text{ g W} \quad 0.9643 \text{ g Cl}/ \text{ g W}$$
$$\text{Compound 4} \quad 53.64 \text{ g Cl}/46.36 \text{ g W} \quad 1.1570 \text{ g Cl}/ \text{ g W}$$

The ratios of each mass of chlorine to the smallest one are

$$0.7715/0.3856 = \quad 2.0008 = \quad 4:2$$
$$0.9643/0.3856 = \quad 2.5008 = \quad 5:2$$
$$1.1570/0.3856 = \quad 3.0005 = \quad 6:2$$

One possible set of formulas is WCl_2, WCl_4, WCl_5, and WCl_6. The same result is confirmed by using a table of relative atomic masses.

1-12 The only products are gaseous N_2 and gaseous H_2. From the formula of the starting compound there are twice as many molecules of H_2 as of N_2 in the products. The law of combining volumes (or, in this case, the law of "uncombining" volumes) then assures that the volume of hydrogen is twice the volume of nitrogen as long as the temperature and pressure remain unchanged. The answer is 27.4 mL.

1-14 The balanced chemical equation for this reaction is:

$$2\,CH_3OH + 3\,O_2 \rightarrow 2\,CO_2 + 4\,H_2O$$

2.0 L of CO_2 and 4.0 L of H_2O are produced from 2.0 L of CH_3OH, according to the law of combining volumes.

1-16 (a) Promethium has an atomic number of 61; the ratio of the number of neutrons to protons in ^{145}Pm is $(145 - 61)/61 = 1.377$.

(b) A neutral atom of Pm has 61 electrons; this +3 ion has 58 electrons.

1-18 The $^{266}_{109}Mt$ atom has 109 protons, 109 electrons, and 157 neutrons.

1-20 In the following the melting points (top line), boiling points (middle line), and densities (bottom line) of the four immediate neighbors of technetium are arrayed as in the periodic table.

$$
\begin{array}{ccc}
& 1244 & \\
& 1962 & \\
& 7.2 & \\
2610 & \boxed{\;Tc\;} & 2310 \\
5560 & & 3900 \\
10.2 & & 12.3 \\
& 3180 & \\
& 5627 & \\
& 20.5 &
\end{array}
$$

We have no reason to treat trends across the table as any more or less strong than trends down the table. Hence, we just average the four melting points to get a predicted melting point, and do the same with the boiling points and the densities to get a predicted boiling point and density. The predictions are 2336°C for melting, 4262°C for boiling, and 12.55 g cm^{-3} for the density. The experimental values are 2172°C for melting, to which the prediction comes reasonably close, 4877°C for boiling, and 11.50 g cm^{-3} for the density.

1-22 The predicted formulas are GeH_4, HF, H_2Te, BiH_3.

1-24 The atomic mass of naturally occuring neon is found by multiplying the isotope fractional abundances by their masses and adding

$$(0.9000 \times 19.99212) + (0.0027 \times 20.99316) + (0.0973 \times 21.99132) = 20.19$$

1-26 This problem resembles problem 1-24, except that the relative mass of one of the five isotopes of Zr is not known, and the relative atomic mass of natural zirconium (91.224) is known. The natural abundance of the isotope of interest is known by difference. It is $[1 - 0.1127 - .1717 - .1733 - 0.0278] = 0.5145$. Let the relative mass of this isotope be x. Then

$$91.224 = 0.5145x + 0.1127 \times 90.9056 + 0.1717 \times 91.9050$$

$$+0.1733 \times 93.9063 + 0.0278 \times 95.9083$$

Solving gives $x = 89.91$.

1-28 The 100 million atoms of fluorine will weigh $18.998403/12.00000$ times the mass of 100 million atoms of ^{12}C. Avogadro's number of ^{12}C atoms weighs exactly 12 g, so 100 million atoms of ^{12}C weighs

$$12.00000 \text{ g} \times \left(\frac{100,000,000}{6.022137 \times 10^{23}} \right) = 1.992648 \times 10^{-15} \text{ g}$$

The mass of the fluorine is 3.154761×10^{-15} g.

1-30 The point of the problem is to assure that students properly handle the nesting of parentheses in chemical formulas when they compute molecular masses and molar masses. The answers: (a) 177.382; (b) 598.156; (c) 254.2; (d) 98.079; (e) 450.446. There are no units in these answers because they are relative masses.

1-32 There are Avogadro's number of gold atoms in a mole of gold, each with a diameter of 2.88×10^{-10}m. The length of the line is $(6.022 \times 10^{23})(2.88 \times 10^{-10} \text{ m}) = 1.734 \times 10^{14}$ m. This is over 1100 times the distance from the earth to the sun.

1-34 Before we can arrange the four samples by mass, we must express the amounts of each in the same unit of mass. In the case of the SF_4, we convert from the given number of moles to grams using the molar mass. In the cases of the Cl_2O_7 and Ar, a preliminary conversion from the given number of particles to chemical amount is necessary; the amount of CH_4 is already in grams. The results are:

$$SF_4 \text{ (115 g)} \quad < \quad CH_4 \text{ (117 g)} \quad < \quad Cl_2O_7 \text{ (264 g)} \quad < \quad Ar \text{ (2770 g)}$$

1-36

$$10.0 \text{ cm}^3 \text{ Au} \times \left(\frac{19.32 \text{ g Au}}{1 \text{ cm}^3 \text{ gold}} \right) \times \left(\frac{1 \text{ troy ounce}}{31.1035 \text{ g Au}} \right) \times \left(\frac{\$400}{1 \text{ troy ounce}} \right)$$

$$= \$2.4846 \times 10^3$$

The cost to three significant figures is \$2480.

1-38 Under these conditions, 415 cm^3 contains

$$(415 \text{ cm}^3)(0.00278 \text{ g cm}^{-3}) = 1.1537 \text{ g} = 0.0185 \text{ mol Si}_2\text{H}_6$$

The number of molecules is Avogadro's number times this:

$$(0.0185 \text{ mol})(6.022 \times 10^{23} \text{ molecules mol}^{-1}) = 1.117 \times 10^{22} \text{ molecules}$$

Because each molecule of Si_2H_6 contains two Si atoms,

$$\text{atoms Si} = 2 \times 1.117 \times 10^{22} = 2.23 \times 10^{22} \text{ atoms Si}$$

1-40 Let us assume a mass of 5.0 ounces (0.14kg) for the baseball. Convert first to SI units:

$$(5.0 \text{ oz}) \left(\frac{1 \text{ lb}}{16 \text{ oz}} \right) \left(\frac{0.4536 \text{ kg}}{1 \text{ lb}} \right) = 0.142 \text{ kg}$$

$$\left(95 \frac{\text{miles}}{\text{hr}} \right) \left(\frac{5280 \text{ ft}}{1 \text{ mile}} \right) \left(\frac{0.3048 \text{ m}}{1 \text{ ft}} \right) \left(\frac{1 \text{ hr}}{3600 \text{ s}} \right) = 42.5 \text{ m s}^{-1}$$

The work done on the ball is equal to its change in kinetic energy, $\frac{1}{2}mv^2 - 0 = \frac{1}{2}mv^2$.

$$\text{Work} = \frac{1}{2}mv^2 = \frac{1}{2}(0.142 \text{ kg})(42.5 \text{ m s}^{-1})^2$$

$$= 128 \text{ kg m}^2 \text{ s}^{-2} = 1.3 \times 10^2 \text{ J}$$

1-42 (a) As suggested in the hint, a good approach to this problem is to look at differences in charge. So, ranking the droplets from least to most charge, we also note the amount the charge differs from the charge on the previous droplet in the list.

Drop	Charge (all ×10^{-19} C)	Difference (all ×10^{-19} C)	Units of Charge (Charge/1.63 × 10^{-19} C)
1	6.563	—	4
2	8.204	1.641	5
3	11.50	3.296	7
4	13.13	1.63	8
5	16.48	3.35	10
6	18.08	1.60	11
7	19.71	1.63	12
8	22.89	3.28	14
9	26.18	3.29	16

The data all differ from the adjacent values in the list by either 1.63×10^{-19} C or twice that value. This suggests that there is a fundamental unit of charge equal to 1.63×10^{-19} C. Dividing this quantity into the observed charges gives the number of electrons on each droplet.

(b) If we divide each observed charge by the integral charge the droplet seems to have, we get these values for the electron charge:

1.6407×10^{-19} C 1.640×10^{-19} C 1.6425×10^{-19} C
1.6408×10^{-19} C 1.648×10^{-19} C 1.635×10^{-19} C
1.643×10^{-19} C 1.6436×10^{-19} C 1.636×10^{-19} C

The average of all these values is 1.641×10^{-19} C.

(c) We have chosen the least possible charge to fit the data. The data could be any fraction (1/2, 1/3, 1/4, etc.) of the value suggested. One could only check the result by a thorough search for droplets with fractional charge.

1-44 Density of neutron star $= \dfrac{\text{mass}}{\text{volume}}$

$$\rho = \frac{6.0 \times 10^{56} \times 1.675 \times 10^{-24} \text{ g}}{(4/3)\pi(20 \times 10^5 \text{ cm})^3} = 3.0 \times 10^{13} \text{ g cm}^{-3}$$

Mass of ^{232}Th nucleus
$= 142(1.675 \times 10^{-24} \text{ kg}) + 90(1.673 \times 10^{-24} \text{ g}) = 3.884 \times 10^{-22} \text{ g}$

$$\text{Density } = \frac{3.884 \times 10^{-22} \text{ g}}{(4/3)\pi(9.1 \times 10^{-13} \text{ cm})^3} = 1.2 \times 10^{14} \text{ g cm}^{-3}$$

This is four times larger than the density of a neutron star.

1-46 Let x be the <u>fractional</u> abundance of ^{85}Rb and $1 - x$ be the abundance of ^{87}Rb. Then

$$x(84.9117) + (1 - x)(86.9092) = 85.4678$$
$$1.4414 = 1.9975x$$
$$x = 0.7216$$

The percentage of ^{85}Rb is 72.16%; ^{87}Rb is 27.84%.

1-48

$$(2.0 \text{ oz}) \left(\frac{1 \text{ lb}}{16 \text{ oz}} \right) \left(\frac{0.4536 \text{ kg}}{1 \text{ lb}} \right) = 0.057 \text{ kg}$$

$$\left(98 \frac{\text{miles}}{\text{hr}} \right) \left(\frac{5280 \text{ ft}}{1 \text{ mile}} \right) \left(\frac{0.3048 \text{ m}}{1 \text{ ft}} \right) \left(\frac{1 \text{ hr}}{3600 \text{ s}} \right) = 43.8 \text{ m s}^{-1}$$

$$\text{Kinetic energy} = \frac{1}{2}mv^2 = \frac{1}{2}(0.057 \text{ kg})(43.8 \text{ m s}^{-1})^2 = 55 \text{ J}$$

No work is done on the building because the wall is not displaced.

Chapter 2

Chemical Equations and Reaction Yields

2-2 The molar mass of acetaminophen, $C_8H_9NO_2$, is 151.165 g mol^{-1}. The percentages are found by dividing this into the mass of each element in one mole of compound, and multiplying by 100%

$$\% \text{ C} \quad \frac{8 \times 12.011 \text{ g mol}^{-1}}{151.165 \text{ g mol}^{-1}} \times 100\% = 63.56\% \text{ C}$$

$$\% \text{ H} \quad \frac{9 \times 1.00794 \text{ g mol}^{-1}}{151.165 \text{ g mol}^{-1}} \times 100\% = 6.001\% \text{ H}$$

$$\% \text{ N} \quad \frac{14.0067 \text{ g mol}^{-1}}{151.165 \text{ g mol}^{-1}} \times 100\% = 9.266\% \text{ N}$$

$$\% \text{ O} \quad \frac{2 \times 15.9994 \text{ g mol}^{-1}}{151.165 \text{ g mol}^{-1}} \times 100\% = 21.17\% \text{ O}$$

2-4 The mass percentage of fluorine in each of the compounds can certainly be calculated and the resulting numbers used to get the required order. It saves work however to <u>estimate</u> the fluorine content of each compound. Thus, HF is certainly the richest possible compound in fluorine by mass because the only other atom is the very light hydrogen atom—there is only 1 unit of non-fluorine mass per fluorine atom. The non-fluorine mass per fluorine atom in C_6HF_5 is about $(6 \times 12)/5 \approx 14$; in BrF it is 79.9; in UF$_6$ it is $238/6 \approx 40$. The desired order is therefore BrF $<$ UF$_6$ $<$ C$_6$HF$_5$ $<$ HF.

2-6 The pharmacist mixes 286 g of one compound with 150 g of another, using water as a mixing agent. After all the water is driven away, the mixture ob-

viously weighs 436 g. The first compound (Na_2CO_3) contains a fixed proportion of carbon as does the second ($C_2H_5NO_2$). The mass of carbon in the
Na_2CO_3 is $(12.011/105.989) \times 286$ g and the mass of carbon in the $C_2H_5NO_2$
is $(2 \times 12.011/75.067) \times 150$ g where the 105.989 and 75.067 are the respective
molecular masses of the compounds. The mass of carbon in the mixture is the
sum of these two masses. It equals 80.411 g. The mass percentage of carbon is
this mass divided by 436 g and multiplied by 100%. It is 18.4%.

2-8 Imagine a sample of bromoform of arbitrary mass, say 100.0 g. The mass of
bromine in this sample is 94.85 g, the mass of hydrogen is 0.40 g, and the mass
of carbon is 4.75 g. Convert each of these masses to chemical amount by dividing
by the molar mass of the element: there are 1.19445 mol of Br, 0.39685 mol of
H, and 0.39547 mol of C. (Nonsignificant figures appear in these intermediate
values for the sake of greater precision in the final result.) The three chemical
amounts stand in the ratio of 3.02 to 1.003 to 1. Within the precision of the
data this ratio is 3 to 1 to 1. The empirical formula is Br_3HC.

2-10 In 100.0000 g of the compound there is 1.6907 g of O and 98.3093 g of Cs. The
chemical amount of oxygen in the 100.00 g of compound is its mass divided
by its molar mass; it equals 0.10567 mol. The chemical amount of cesium is
0.73969 mol. The ratio of the chemical amounts is 7.000 to 1, making the
empirical formula Cs_7O.

2-12 The empirical formulas of the five compounds are

Compound	Formula
A	CO_2
B	CO
C	C_4O_3
D	C_3O_2
E	C_5O_2

All five of these compositions exist. The first two are commonplace. The third
is mellitic anhydride and has the molecular formula $C_{12}O_9$. The last two have
molecular formulas identical to their empirical formulas and are "suboxides" of
carbon.

2-14 (a)
$$\text{moles Ca} = \text{moles CaO} = \frac{2.389 \text{ g}}{56.0774 \text{ g mol}^{-1}} = 0.04260 \text{ mol}$$

$$\text{moles C} = \text{moles CO}_2 = \frac{1.876 \text{ g}}{44.010 \text{ g mol}^{-1}} = 0.04263 \text{ mol}$$

$$\text{moles N} = \text{moles NO}_2 = \frac{3.921 \text{ g}}{46.0055 \text{ g mol}^{-1}} = 0.08523 \text{ mol}$$

$$\frac{\text{mol N}}{\text{mol Ca}} = \frac{0.08523}{0.04260} = 2.001 \approx 2$$

The empirical formula is $CaCN_2$.

2-16 Because a given volume contains the same number of molecules, the ratio of molecular masses equals the ratio of densities:

$$\frac{\text{molecular mass of P}_n}{\text{molecular mass of N}_2} = \frac{2.7}{0.62}$$

$$\text{molecular mass of P}_n = \left(\frac{2.7}{0.62}\right)(28.014) = 122$$

Because the atomic mass of P is 30.97, n must be

$$n = \frac{122}{30.97} = 3.94 \approx 4$$

There are four P atoms per molecule under these conditions.

2-18 (a) Molecular mass $= (2.53)(28.013) = 70.9$

(b) $\dfrac{8.21 \text{ g}}{70.9 \text{ g mol}^{-1}} = 0.1158$ mol compound

$\dfrac{1.62 \text{ g}}{14.007 \text{ g mol}^{-1}} = 0.1157$ mol N atoms

Each molecule has one N atom

(c) $71.0 - 14.0 = 57.0$

(d) $\frac{57.0}{2} = 28.5$, $\frac{57.0}{3} = 19.0$, $\frac{57.0}{4} = 14.3$, and so forth.

Best choice is F (atomic mass 19.0).

(e) NF_3

2-20 (a) $2\,Al + 6\,HCl \rightarrow 2\,AlCl_3 + 3\,H_2$

(b) $4\,NH_3 + 5\,O_2 \rightarrow 4\,NO + 6\,H_2O$

(c) $2\,Fe + O_2 + 2\,H_2O \rightarrow 2\,Fe(OH)_2$

(d) $2\,HSbCl_4 + 3\,H_2S \rightarrow Sb_2S_3 + 8\,HCl$

(e) $2\,Al + Cr_2O_3 \rightarrow Al_2O_3 + 2\,Cr$

(f) $XeF_4 + 2\,H_2O \rightarrow Xe + O_2 + 4\,HF$

(g) $(NH_4)_2Cr_2O_7 \rightarrow N_2 + Cr_2O_3 + 4\,H_2O$

(h) $NaBH_4 + 2\,H_2O \rightarrow NaBO_2 + 4\,H_2$

2-22 (a)

$$\left(\frac{1.000 \text{ g CaCO}_3}{100.09 \text{ g mol}^{-1}}\right)\left(\frac{1 \text{ mol Ca(OH)}_2}{1 \text{ mol CaCO}_3}\right) \times 74.094 \text{ g mol}^{-1} = 0.7403 \text{ g Ca(OH)}_2$$

(b)

$$\left(\frac{1.000 \text{ g C}_3\text{H}_8}{44.096 \text{ g mol}^{-1}}\right)\left(\frac{3 \text{ mol CO}_2}{1 \text{ mol C}_3\text{H}_8}\right) \times 44.010 \text{ g mol}^{-1} = 2.994 \text{ g CO}_2$$

(c)

$$\left(\frac{1.000 \text{ g MgNH}_4\text{PO}_4}{137.315 \text{ g mol}^{-1}}\right)\left(\frac{1 \text{ mol Mg}_2\text{P}_2\text{O}_7}{2 \text{ mol MgNH}_4\text{PO}_4}\right) \times 222.55 \text{ g mol}^{-1} = 0.8104 \text{ g Mg}_2\text{P}_2\text{O}_7$$

2-24 Whatever the reaction or series of reactions that gives $Pt_2C_{10}H_{18}N_2S_2O_6$, the chemical amount that forms cannot exceed half the chemical amount of platinum because the compound contains two Pt per molecule. Thus:

$$1.406 \text{ g Pt} \times \left(\frac{1 \text{ mol Pt}}{195.08 \text{ g Pt}}\right) \times \left(\frac{1 \text{ mol Pt}_2\text{C}_{10}\text{H}_{18}\text{N}_2\text{S}_2\text{O}_6}{2 \text{ mol Pt}}\right)$$

$$\times \left(\frac{716.55 \text{ g Pt}_2\text{C}_{10}\text{H}_{18}\text{N}_2\text{S}_2\text{O}_6}{1 \text{ mol Pt}_2\text{C}_{10}\text{H}_{18}\text{N}_2\text{S}_2\text{O}_6}\right) = 2.582 \text{ g Pt}_2\text{C}_{10}\text{H}_{18}\text{N}_2\text{S}_2\text{O}_6$$

2-26 Use the equation $Si_4H_{10} + 13/2\,O_2 \rightarrow 4\,SiO_2 + 5\,H_2O$:

$$(25.0 \text{ cm}^3)(0.825 \text{ g cm}^{-3}) = 20.625 \text{ g Si}_4\text{H}_{10}$$

$$\left(\frac{20.625 \text{ g Si}_4\text{H}_{10}}{122.42 \text{ g mol}^{-1}}\right) \times \left(\frac{4 \text{ mol SiO}_2}{1 \text{ mol Si}_4\text{H}_{10}}\right) \times \left(\frac{60.0843 \text{ g SiO}_2}{1 \text{ mol SiO}_2}\right) = 40.5 \text{ g SiO}_2$$

2-28

$$67.2 \text{ g S} \times \left(\frac{1 \text{ mol S}}{32.066 \text{ g S}}\right) \times \left(\frac{1 \text{ mol CS}_2}{4 \text{ mol S}}\right) \times \left(\frac{76.143 \text{ g CS}_2}{1 \text{ mol CS}_2}\right) = 39.9 \text{ g CS}_2$$

2-30

$$\frac{69.8 \text{ g P}_4}{123.895 \text{ g mol}^{-1}} = 0.5634 \text{ mol P}_4$$

$$(0.5634 \text{ mol P}_4)\left(\frac{2 \text{ mol Ca}_3(\text{PO}_4)_2}{1 \text{ mol P}_4}\right) = 1.1268 \text{ mol Ca}_3(\text{PO}_4)_2$$

$$(1.1268 \text{ mol})(310.18 \text{ g mol}^{-1}) = 349 \text{ g Ca}_3(\text{PO}_4)_2$$

$$(0.5634 \text{ mol P}_4)\left(\frac{6 \text{ mol CaSiO}_3}{1 \text{ mol P}_4}\right) = 3.380 \text{ mol CaSiO}_3$$

$$(3.380 \text{ mol})(116.16 \text{ g mol}^{-1}) = 393 \text{ g CaSiO}_3$$

2-32 Let $\mathcal{M}$ be the molar mass of A. Then $\mathcal{M} + 3(126.90)$ is the molar mass of AI_3.

$$\text{moles AI}_3 = \frac{0.8000 \text{ g}}{\mathcal{M} + 3(126.90) \text{ g mol}^{-1}} = \text{moles ACl}_3$$

$$\text{grams ACl}_3 = \frac{0.8000 \text{ g}}{\mathcal{M} + 3(126.90) \text{ g mol}^{-1}} \times (\mathcal{M} + 3(35.453) \text{ g mol}^{-1} = 0.3776 \text{ g}$$

$$0.8000[\mathcal{M} + 106.36] = 0.3776[\mathcal{M} + 380.7]$$

Solving for $\mathcal{M}$ gives $\mathcal{M} = 138.9 \text{ g mol}^{-1}$

(b) The element is lanthanum, La.

2-34

$$\frac{0.738 \text{ g}}{2.016 \text{ g mol}^{-1}} = 0.366 \text{ mol H}_2$$

Suppose the mass of iron originally was x g. Then

$$\text{moles Fe} = \frac{x \text{ g Fe}}{55.847 \text{ g mol}^{-1}}$$

$$\text{moles Al} = \frac{(9.62 - x) \text{ g Al}}{26.982 \text{ g mol}^{-1}}$$

Hydrogen (0.366 mol) is produced by the reaction of Fe and Al. Adding the two contributions gives

$$\frac{x \text{ g Fe}}{55.847 \text{ g mol}^{-1}}\left(\frac{1 \text{ mol H}_2}{1 \text{ mol Fe}}\right) + \frac{(9.62 - x) \text{ g Al}}{26.982 \text{ g mol}^{-1}}\left(\frac{3 \text{ mol H}_2}{2 \text{ mol Al}}\right) = 0.366 \text{ mol H}_2$$

$$-0.03769x + 0.5348 = 0.366$$

$$x = 4.48 \text{ g Fe}$$

2-36 (a) $NH_3 + CH_4 \rightarrow HCN + 3\,H_2$

(b)

$$\frac{500.0 \text{ g CH}_4}{16.043 \text{ g mol}^{-1}} = 31.17 \text{ mol CH}_4$$

$$\frac{200.0 \text{ g NH}_3}{17.031 \text{ g mol}^{-1}} = 11.74 \text{ mol NH}_3$$

NH_3 is the limiting reactant. After the reaction there will be

$$11.74 \text{ mol NH}_3 \left(\frac{1 \text{ mol HCN}}{1 \text{ mol NH}_3}\right) \times 27.026 \text{ g mol}^{-1} = 317.3 \text{ g HCN}$$

$$11.74 \text{ mol NH}_3 \left(\frac{3 \text{ mol H}_2}{1 \text{ mol NH}_3}\right) \times 2.0158 \text{ g mol}^{-1} = 71.0 \text{ g H}_2$$

$$(31.17 - 11.74 \text{ mol CH}_4)(16.043 \text{ g mol}^{-1}) = 311.7 \text{ g CH}_4$$

2-38 The theoretical yield of $TiCl_4$ is

$$\left(\frac{7390 \text{ g TiO}_2}{79.88 \text{ g mol}^{-1}}\right) \times \left(\frac{1 \text{ mol TiCl}_4}{1 \text{ mol TiO}_2}\right) \times \left(189.69 \text{ g mol}^{-1}\right) = 1.755 \times 10^4 \text{ g}$$

$$= 17.55 \text{ kg TiCl}_4$$

The percentage yield is

$$\left(\frac{14.24 \text{ kg}}{17.55 \text{ kg}}\right) \times 100\% = 81.1\%$$

2-40 The mass percentages in the two oxides are given to six significant figures; we carry out the calculation to equally high precision. There is 0.43132 mol of W in a 100.000 g sample of the white oxide, and 1.29395 mol of O in the same sample. The ratio of these chemical amounts is 3.0000—the empirical formula is WO_3. In the blue oxide the chemical amounts in a 100.000 g sample are 0.43975 mol W and 1.19709 mol of O. The ratio of these two is 2.72223. This turns out to equal the ratio of 49 to 18, within the precision of the data. Hence the formula $W_{18}O_{49}$ is a correct answer. The blue oxide is really a nonstoichiometric compound, however.

2-42 Consider a 100.00-g sample of this binary compound. It contains 78.06 g of Ni and 21.94 g of O. This is $78.06 \text{ g}/58.69 \text{ g mol}^{-1} = 1.330$ mol of Ni and $21.94 \text{ g}/15.9994 \text{ g mol}^{-1} = 1.371$ mol of O. The ratio of these two chemical amounts is 1.031 to 1. If the data are truly precise to four significant figures, the compound is almost certainly a nonstoichiometric compound. The "almost" appears because "$Ni_{1000}O_{1031}$" is a conceivable stoichiometric formulation. These subscript are whole numbers, but they are hardly small whole numbers.

2-44

$$\text{Chemical amount of B} = \frac{0.664 \text{ g B}}{10.811 \text{ g mol}^{-1}} = 0.0614 \text{ mol B}$$

Mass of Cl in original compound $= 2.842 - 0.664 \text{ g} = 2.178 \text{ g Cl}$

$$\text{Chemical amount of Cl} = \frac{2.178 \text{ g Cl}}{35.453 \text{ g mol}^{-1}} = 0.0614 \text{ mol Cl}$$

$$\frac{0.0614 \text{ mol B}}{0.0614 \text{ mol Cl}} = \frac{1 \text{ mol B}}{1 \text{ mol Cl}} \text{ so the empirical formula is BCl}$$

$$\text{Chemical amount of Cl}_2 \text{ produced} = \left(\frac{1 \text{ mol Cl}_2}{2 \text{ mol Cl}}\right) \times (0.0614 \text{ mol Cl})$$

$$= 0.0307 \text{ mol Cl}_2$$

From Avogadro's hypothesis, the chemical amount of compound in the original 0.153 L must be

$$\left(\frac{0.153 \text{ L compound}}{0.688 \text{ L Cl}_2}\right) \times 0.0307 \text{ mol Cl}_2 = 0.00683 \text{ mol compound}$$

Its approximate molar mass is

$$\frac{2.842 \text{ g}}{0.00683 \text{ mol}} = 416.3 \text{ g mol}^{-1}$$

The molar mass of the empirical formula, BCl, is $46.264 \text{ g mol}^{-1}$.

$\frac{416.3}{46.264} = 9.00$, so the molecular formula must be B_9Cl_9.

2-46 The balanced chemical equation reads

$$C_{14}H_{18}N_2O_5 + 2 \text{ H}_2O \rightarrow C_4H_7NO_4 + CH_3OH + \text{phenylalanine}$$

By balancing the chemical amounts of C, H, N and O on both sides, we determine the molecular formula of phenylalanine to be $C_9H_{12}NO_2$.

2-48 If we assume that all the carbon is evolved as CH_4, then 1 mol Al_4C_3 generates 3 mol CH_4

$$\text{moles } CH_4 = \left(\frac{63.2 \text{ g } Al_4C_3}{143.96 \text{ g mol}^{-1}}\right)\left(\frac{3 \text{ mol } CH_4}{1 \text{ mol } Al_4C_3}\right) = 1.317 \text{ mol}$$

$$(1.317 \text{ mol})(16.043 \text{ g mol}^{-1}) = 21.1 \text{ g}$$

2-50

$$\frac{0.211 \text{ g } CO_2}{44.010 \text{ g mol}^{-1}} = 0.004794 \text{ mol } CO_2$$

Let x = mass of $SrCO_3$; $0.800 - x$ = mass of $BaCO_3$.

$$\text{moles } SrCO_3 = \frac{x \text{ g}}{147.63 \text{ g mol}^{-1}}$$

$$\text{moles } BaCO_3 = \frac{0.800 - x \text{ g}}{197.34 \text{ g mol}^{-1}}$$

Because each mole of $SrCO_3$ and $BaCO_3$ gives one mole of CO_2,

$$\frac{x}{147.63} + \frac{0.800 - x}{197.34} = 0.004794$$

$$x = 0.434 \text{ g } SrCO_3$$

$$\% \text{ } SrCO_3 = \frac{0.434}{0.800} \times 100\% = 54.2\%$$

2-52 Suppose your gas mileage is 20 miles per gallon. Then reducing driving by 20 miles per week would save 1 gallon per week, or 52 gallons per year. Converted to the metric system this is

$$52 \text{ gallons} \times \left(\frac{3.785 \text{ L}}{1 \text{ gallon}}\right) = 200 \text{ L gasoline}$$

Take the density of gasoline to be that of water, which is 1 g cm^{-3}, or 1 kg L^{-1}. This is then 200 kg of gasoline. Using the chemical formula C_8H_{18}, it is

$$\frac{200 \times 10^3 \text{ g } C_8H_{18}}{114.2 \text{ g mol}^{-1}} = 1.75 \times 10^3 \text{ mol } C_8H_{18}$$

Because 1 mol C_8H_{18} generates 8 mol CO_2, the chemical amount of CO_2 produced is 8 times this, or 1.4×10^4 mol CO_2. Its mass is

$$\left(1.4 \times 10^4 \text{ mol}\right) \times \left(44 \text{ g mol}^{-1}\right) = 6.2 \times 10^5 \text{ g} = 620 \text{ kg}$$

Converting this to pounds gives

$$620 \text{ kg} \left(\frac{1 \text{ pound}}{0.4536 \text{ kg}} \right) = 1400 \text{ pounds}$$

This is on the order of 1000 pounds, consistent with the statement in the problem.

2-54 The theoretical yield of the $KClO_4$ must be computed and then compared to the observed yield of 3.00 g. First we balance the chemical equation:

$$4 \text{ KClO}_3(s) \rightarrow 3 \text{ KClO}_4(s) + \text{KCl}(s)$$

$$4.00 \text{ g KClO}_3 \times \left(\frac{1 \text{ mol KClO}_3}{122.549 \text{ g KClO}_3} \right) \times \left(\frac{3 \text{ mol KClO}_4}{4 \text{ mol KClO}_3} \right)$$

$$\times \left(\frac{138.549 \text{ g KClO}_4}{1 \text{ mol KClO}_4} \right) = 3.392 \text{ g KClO}_4$$

The comparison is performed by dividing the actual by the theoretical yield and multiplying by 100%. The result is 88.5%.

2-56 The yield of the product is less than the theoretical, and this fact must be reckoned with. A very good way is to use the percentage yield to construct an additional unit-factor (the second factor in the following):

$$125 \text{ g Si}_3\text{N}_4 \text{ isolated} \times \left(\frac{100 \text{ g Si}_3\text{N}_4 \text{ formed}}{95.0 \text{ g Si}_3\text{N}_4 \text{ isolated}} \right) \times \left(\frac{1 \text{ mol Si}_3\text{N}_4}{148.286 \text{ g Si}_3\text{N}_4 \text{ formed}} \right)$$

$$\times \left(\frac{3 \text{ mol Si}}{1 \text{ mol Si}_3\text{N}_4} \right) \times \left(\frac{28.086 \text{ g Si}}{1 \text{ mol Si}} \right) = 74.8 \text{ g Si}$$

Chapter 3

Chemical Bonding: The Classical Description

3-2 (a) $(Ra)^{2+}$ 86 electrons (0 valence, 86 core)

(b) $:\overset{\cdot\cdot}{\underset{\cdot\cdot}{Br}}\cdot$ 35 electrons (7 valence, 28 core)

(c) $(:\overset{\cdot\cdot}{\underset{\cdot\cdot}{Bi}}\cdot)^{2-}$ 85 electrons (7 valence, 78 core)

(d) $(\cdot Ga\cdot)^{+}$ 30 electrons (2 valence, 28 core)

3-4 (a) $GaBr_3$, gallium bromide

$$\cdot \overset{\cdot}{Ga} \cdot + 3 : \overset{\cdot\cdot}{\underset{\cdot\cdot}{Br}}\cdot \rightarrow (Ga^{3+})(: \overset{\cdot\cdot}{\underset{\cdot\cdot}{Br}} :^{-})_3$$

(b) SrPo, strontium polonide

$$\cdot Sr \cdot + : \overset{\cdot}{\underset{\cdot}{Po}} : \rightarrow (Sr^{2+})(: \overset{\cdot\cdot}{\underset{\cdot\cdot}{Po}} :^{2-})$$

(c) MgI_2, magnesium iodide

$$\cdot Mg \cdot + 2 : \overset{\cdot\cdot}{\underset{\cdot\cdot}{I}}\cdot \rightarrow (Mg^{2+})(: \overset{\cdot\cdot}{\underset{\cdot\cdot}{I}} :^{-})_2$$

(d) Li_2Se, lithium selenide

$$Li \cdot + : \overset{\cdot}{\underset{\cdot}{Se}} : \rightarrow (Li^{+})_2(: \overset{\cdot\cdot}{\underset{\cdot\cdot}{Se}} :^{2-})$$

3-6 (a) potassium nitrite   (b) strontium permanganate   (c) magnesium dichromate
(d) sodium dihydrogen phosphate   (e) barium chloride   (f) sodium chlorate

3-8 (a) Cs_2SO_3 (b) $Sr(SCN)_2$ (c) LiH   (d) Na_2O_2 (e) $(NH_4)_2Cr_2O_7$ (f) $RbHSO_4$

3-10 The formula is $NH_4H_2PO_4$, and the systematic name is ammonium dihydrogen phosphate.

3-12 The overall process

$$RbF \rightarrow Rb + F$$

is the sum of three steps:

$$
\begin{array}{rcll}
RbF(g) & \rightarrow & Rb^+(g) + F^-(g) & \Delta E_1 \\
Rb^+(g) + e^- & \rightarrow & Rb(g) & \Delta E = -IE_1(Rb) \\
F^-(g) & \rightarrow & F(g) + e^- & \Delta E = EA(F) \\
\hline
\text{Total}\quad RbF(g) & \rightarrow & Rb(g) + F(g) & \Delta E_{\text{total}}
\end{array}
$$

The energy change for the first step is found by assuming the RbF to be ionic. The energy to to separate an Rb^+ ion and an F^- ion, initially 2.274 Å apart, is

$$\Delta E_1 = \frac{e^2}{4\pi\epsilon_o R} = \frac{(1.602 \times 10^{-19}\ \text{C})^2}{4\pi(8.854 \times 10^{-12}\ \text{C}^2\ \text{J}^{-1}\ \text{m}^{-1})(2.274 \times 10^{-10}\ \text{m})}$$

$$= 1.014 \times 10^{-18}\ \text{J} = 611\ \text{kJ mol}^{-1}$$

Then the dissociation energy of RbF to Rb and F is

$$\Delta E = \Delta E_1 - IE_1(Rb) + EA(F) = 611 - 403 + 328 = 536\ \text{kJ mol}^{-1}$$

This is close to the experimental value of 489 kJ mol^{-1}.

3-14 The order of bond length is ClF < BrCl < IBr. The smallest bond enthalpy should be for the longest bond, the one in IBr.

3-16 (a) The formal charge on the central chlorine atom is +3; all four oxygen atoms have formal charges of −1.

(b) The formal charge of the central sulfur in the Lewis diagram for sulfur dioxide is +1; the left-hand oxygen has a formal charge of −1, and the right-hand oxygen has a formal charge of 0.

(c) The formal charge on the central bromine in this representation of the bonding in the bromite ion is +1. The two oxygens both have formal charges of −1.

(d) The formal charge on the central nitrogen in this Lewis diagram for nitrate ion is +1. The left-hand oxygen has a formal charge of 0, and the other two oxygen atoms have formal charges of −1.

3-18 In the Cl–Cl–O structure, the central Cl atom has formal charge +1 and the O atom has formal charge −1. In the Cl–O–Cl structure, each atom has formal charge zero. The second structure is favored because it shows less separation of formal charge.

3-20 (a) The "Z" represents a Group V element, such as N.

(b) The unknown main group element is a Group VII element, such as Cl.

(c) The unknown main group element is a Group VI element, such as S.

(d) The "Z" is a Group V element, such as N.

3-22 The Lewis dot structures are:

(a)   (b)   (c)   (d)

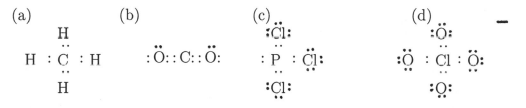

3-24

The C–H bond lengths are estimated to be 1.10×10^{-10} m, the C–C bond length is 1.54×10^{-10} m, the C=O bond length is 1.20×10^{-10} m, the C–O bond length is 1.43×10^{-10} m, and the O–H bond length is 0.96×10^{-10} m.

3-26 Each of the four phosphorus atoms has a lone pair of electrons. This accounts for 8 electrons. Each of the six dotted lines is replaced by a pair of electrons. This uses 12 electrons. Thus 20 electrons are used. These 20 electrons are all the valence electrons furnished by the four P atoms in P_4.

3-28 (a)

: H: N::: C:

(b)

⊖ :S̈: C::: N: ↔   S̈:: C:: N̈: ⊖

(c)

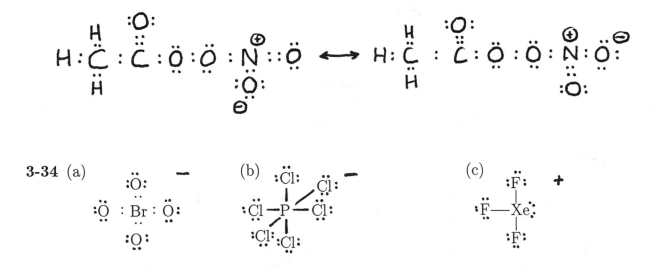

Ḣ   ⊕   Ḣ   ⊕
C̈ :: N :: N̈ ⊖   ↔   ⊖ :C̈ : N ⁝⁝⁝ N̈
Ḣ   Ḣ

3-30

:Ö:⊖   :O: :Ö:⊖
Ö:: C : Ö:⊖   ↔   ⊖Ö : C : Ö:⊖   ↔   ⊖Ö : C:: Ö

The carbon-oxygen bond lengths should fall between the values for a double bond (1.20×10^{-10} m) and for a single bond (1.43×10^{-10} m).

3-32 The Lewis structure is

        :O: :O:
 H   ⁝⁝   H   ⁝⁝
H:C̈:C:Ö:Ö:N⊕::Ö   ⟷   H:C̈:C:Ö:Ö:N⊕:Ö:⊖
 H   :Ö: H   ::Ö:
        ⊖   :O:

3-34 (a)   (b)   (c)
        :Ö: −   :Cl̈:   Cl̈: −   :F̈:   +
   :Ö : Br : Ö: :C̈l—P—C̈l: :F̈—Xe:
        :Q: :Cl̈: Cl̈:   :F̈:

3-36 (a) $MgBr_2$ is more ionic than PBr_3 and should have a higher boiling point.

(b) SrO should boil higher than OsO_4 for the same reason.

(c) Al_2O_3 should boil higher than Cl_2O.

3-38

$$\delta = \frac{(0.2082 \text{ Å debye}^{-1})}{R}\mu = \frac{0.2082 \text{ Å D}^{-1}}{0.980 \text{ Å}}(1.66 \text{ D}) = 0.353 \text{ for OH } (35.3\%)$$

$$\delta = \frac{0.2082 \text{ Å D}^{-1}}{1.131 \text{ Å}}(1.46 \text{ D}) = 0.269 \text{ for CH } (26.9\%)$$

$$\delta = \frac{0.2082 \text{ Å D}^{-1}}{1.175 \text{ Å}}(1.45 \text{ D}) = 0.257 \text{ for CN } (25.7\%)$$

$$\delta = 0 \text{ for } C_2$$

3-40 The third and fourth columns of the following table present the comparison. Note that the absolute value of Δ must always be substituted in the equation.

Compound	Δ	$16\Delta + 3.5\Delta^2$	Expt. Ionic Character
ClF	0.82	15	11%
BrF	1.02	20	15%
BrCl	0.20	3.3	5.6%
ICl	0.50	8.9	5.8%
IBr	0.30	5.1	10%

3-42 (a) Phosphorus trifluoride has a central P with SN 4. The molecule is trigonal pyramidal, like NH_3 (text Figure 14-17).

(b) Sulfuryl chloride has a central S with SN 4. The molecule is close to tetrahedral, but somewhat distorted because of the different steric requirements of the O's and Cl's.

(c) The PF_6^- anion has a central P with SN 6. The anion is octahedral.

(d) The ClO_2^- anion has a central Cl with SN 4. The anion is bent.

(e) Germanium hydride has a central Ge with SN 4. It is tetrahedral.

3-44 (a) In TeH_2, the central Te has SN 4. This molecule is bent.

(b) In AsF_3, the central As has SN 4. The molecule is a trigonal pyramid in which the F—As—F is distorted to somewhat less than the tetrahedral value of 109.5°.

(c) In PCl_4^+, the central P has SN 4. The molecular ion is an undistorted tetrahedron of Cl atoms about the central P.

(d) In XeF_5^+, the central Xe has SN 6. The molecular ion is a square pyramid with the four angles F_{eq}—Xe—F_{ax} distorted to less than 90°.

3-46 There are many possible answers for each part. Examples: (a) ClO_4^- (b) CO_2
(c) SbF_6^- (d) ClO_3^-

3-48 (a) Polar. (b) Polar. (c) Nonpolar. (d) Polar. (e) Nonpolar.

3-50 (a) Using VSEPR concepts, the $GaCl_4^-$ ion has a central Ga with steric number 4; it would be tetrahedral. The $SbCl_4^-$ ion has a central Sb with SN 5. It would have a seesaw geometry (text Figure 14-8b) since one of the five electron pairs is a lone pair.

(b) The $SbCl_2^+$ ion, in which the central Sb has SN 3, is a bent molecular ion, and the $GaCl_2^+$ ion, in which the central Ga has SN 2, is linear. It follows that the formulation $(SbCl_2^+)(GaCl_4^-)$ is more likely correct.

3-52 (a) The observation of a nonzero dipole moment for O_3 rules out a symmetrical linear geometry for the molecule. The molecule of ozone could be linear if the two O-to-O bonds were of different lengths, or if the molecule were bent.

(b) VSEPR assigns a steric number of 3 to the central O and predicts that the molecule of ozone is bent.

3-54 $\overset{-3}{N}\overset{+1}{H_4}\overset{+5}{N}\overset{-2}{O_3}$ $\overset{+2}{Ca}\overset{+2}{Mg}\overset{+4}{Si}\overset{-2}{O_4}$

$\overset{+2}{Fe}(\overset{+2}{C}\overset{-3}{N})_6^{4-}$ $\overset{-3}{B_2}\overset{+1}{H_6}$

$\overset{+2}{Ba}\overset{-1}{H_2}$ $\overset{+2}{Pb}\overset{-1}{Cl_2}$

$\overset{+2}{Cu_2}\overset{-2}{O}(\overset{+6}{S}\overset{-2}{O_4})$ $(\overset{+5/2}{S_4}\overset{-2}{O_6})^{2-}$

The choices for C and N in $Fe(CN)_6^{4-}$ were somewhat arbitrary; other choices are possible.

3-56 (a) La_2S_3 (b) Cs_2SO_4 (c) N_2O_3
(d) IF_5 (e) $Cr_2(SO_4)_3$ (f) $KMnO_4$

3-58 (a) magnesium silicate

(b) iron(II) hydroxide; iron (III) hydroxide

(c) diarsenic pentaoxide or arsenic(V) oxide

(d) ammonium hydrogen phosphate

(e) selenium hexafluoride

(f) mercury(I) sulfate

3-60 (a) In the Lewis structure on the left, the hydrogens and the carbon have formal charges of 0; the sulfur has a formal charge of +1, and the oxygen has a formal charge of -1. In the structure on the right, the hydrogens and the oxygen have zero formal charges; the carbon has a -1 formal charge and the sulfur has a $+1$ formal charge.

(b) The only way to draw a Lewis structure for the molecule of sulfine in which all atoms have formal charges of zero is to violate the octet rule for the sulfur atom (which would have two double bonds and see a total of ten valence electrons).

3-62 (a)

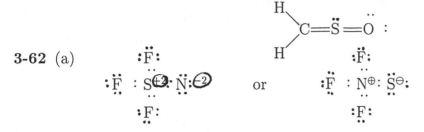

or

(b) The central N is preferred because it gives a smaller separation of formal charge.

3-64 (a)

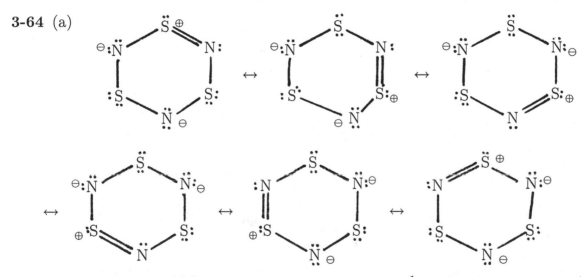

(b) Shown above. (c) Each S atom has a charge of $+\frac{1}{3}$, and each N atom of $-\frac{2}{3}$.

(d) Total charge $= 3(-0.375) + 3(+0.041) = -1.002 \approx -1$

3-66 (a) $^{\oplus}\ddot{\underset{..}{C}}l = \overset{\ominus 2}{Be} = \ddot{\underset{..}{C}}l^{\oplus}$

(b) $\overset{..}{:}\ddot{\underset{..}{C}}l - Be - \ddot{\underset{..}{C}}l\overset{..}{:}$

3-68 $:\ddot{\underset{..}{F}} - \ddot{\underset{..}{S}} - \ddot{\underset{..}{F}}:$

$$
\begin{array}{c}
:\ddot{\text{F}}:\\
|\\
:\ddot{\text{F}}\!-\!\ddot{\text{S}}\!\cdot\!-\!\ddot{\text{S}}\!-\!\ddot{\text{F}}:\\
|\\
:\ddot{\text{F}}:
\end{array}
$$

Valence expansion is necessary for S_2F_4.

3-70 (a) The ΔE of the first reaction is the ionization energy IE_1 of Na(g) (multiplied by one mole) added to the negative of the EA (electron affinity) of I (also multiplied by 1 mol). Taking the values from Appendix F gives $\Delta E = 495.8 + (-295.2) = 200.6$ kJ. The ΔE of the second reaction is the IE of 1 mol of I added to the negative of the electron affinity of 1 mol of Na. It is $\Delta E = 1008.4 + (-52.867) = 955.5$ kJ.

(b) Similar combination of the ionization energies and electron affinities of K and Cl gives ΔE of the first reaction as 69.8 kJ and ΔE of the second reaction as 1202.7 kJ. Even if Na^-I^+ or K^-Cl^+ were to form, the reactions transferring electrons to form Na^+I^- or K^+Cl^- would be strongly favored energetically and would occur quickly.

3-72 For KCl(g)

$$
\mu = 10.3\text{D}; \quad R = 2.67 \text{ Å}
$$

If the KCl(g) molecule were completely ionic, its dipole moment would be

$$
\mu = \frac{2.67 \times 10^{-10} \text{ m} \times 1.602 \times 10^{-19} \text{ C}}{3.336 \times 10^{-30} \text{ C m D}^{-1}} = 12.82 \text{ D}
$$

As it is, the degree of ionicity is not unity, but

$$
\delta = \frac{10.3 \text{ D}}{12.82 \text{ D}} = 0.803 \ (80.3\%)
$$

3-74 The steric number of the central Xe in XeF_2 is 5; the molecule is linear. In XeF_4, the SN of the central Xe is 6; the molecule is square planar. In XeO_3, the SN of the central Xe is 4; the molecule is pyramidal (like ammonia). In XeO_4, the SN of the central Xe is 4; the molecule is tetrahedral. In H_4XeO_6, the 6 O's are bonded to a central Xe that has SN 6. The molecule is octahedral with H's on four of the six oxygen atoms. In $XeOF_4$, the central Xe has SN 6. The molecule consists of a square pyramid surrounding the Xe with the 4 F's at the corners of the base and the O at the apex (this allows the greatest distance between the lone pair and the O).

3-76 The Lewis dot structures are:

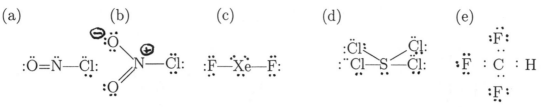

The NOCl molecule is bent and polar; the O_2NCl molecule is (nearly) trigonal about the N and polar; the XeF_2 molecule is linear and non-polar; the SCl_4 molecule has a seesaw geometry and is polar; the CHF_3 molecule is (nearly) tetrahedral and polar.

3-78 From Table 14-6, the order of the bond in C_2 should be 2, the same as that in ethylene, C_2H_4. The bond length should be close to 1.34 Å. The observed value (1.31 Å) is consistent with this prediction, being only slightly smaller.

3-80 $2 \overset{+1\ +7\ -2}{KMnO_4} \rightarrow \overset{+1\ +6\ -2}{K_2MnO_4} + \overset{+4\ -2}{MnO_2} + \overset{0}{O_2}$

Of the two Mn atoms, one gains one electron, and the other gains three, for a total of four. Two oxygen atoms each give up two electrons to provide these four electrons.

3-82 Hydrogen forms a +1 ion in many of its compounds, just like the metals of Group I. The +1 oxidation state in fact predominates in its chemistry. On the other hand, it is a gas not a metal at room conditions like the elements of Group VII). Also, in some situations (metal hydrides) it does form a −1 ion like the elements of Group VII.

3-84 Assume we have 100.00 g of compound and use the procedure of Chapter 1 to find the chemical amounts of each of the elements by dividing masses by molar masses:

$$\frac{48.46 \text{ g O}}{15.9994 \text{ g mol}^{-1}} = 3.029 \text{ mol O}$$

$$\frac{23.45 \text{ g P}}{30.97376 \text{ g mol}^{-1}} = 0.7571 \text{ mol P}$$

$$\frac{21.21 \text{ g N}}{14.0067 \text{ g mol}^{-1}} = 1.514 \text{ mol N}$$

$$\frac{6.87 \text{ g H}}{1.00794 \text{ g mol}^{-1}} = 6.816 \text{ mol H}$$

Divide each by the smallest one to find the formula $PO_4N_2H_9$, which is $(NH_4)_2HPO_4$, or ammonium hydrogen phsophate. The Lewis structures of the two ions are:

3-86 (a) The Lewis structures are $:\ddot{N}=\ddot{S}-\ddot{F}:$ and $:\ddot{S}=\ddot{N}-\ddot{F}:$ and $:\ddot{N}=\ddot{F}-\ddot{S}:$

In the first, the formal charges are (from left to right) -1, $+1$, and 0. In the second, all three atoms have zero formal charge. In the third, the formal charges are -1, $+2$, and -1.

(b) The structure with the least separation of formal charge has a central N. This is not consistent with the observation of a central S atom.

(c) The electronegativity of N exceeds that of S; that is, N has a greater tendency than S to accept electrons in a chemical bond. This helps explain why the observed structure corresponds to a formal build-up of negative charge on the N and formal build-up of positive charge on the S. Also, the two most electronegative atoms (N and F) are separated, reducing electron-electron repulsions.

Chapter 4

The Gaseous State

4-2 The products of this reaction are easily predicted in the context; the balanced equation is: $NH_4CO_2NH_2(s) \rightarrow 2\,NH_3(g) + CO_2(g)$. When predicting the products of a reaction, look for small stable molecules (like those of water, carbon dioxide or ammonia) that might be split off by larger molecules.

4-4 The acidification of aqueous solutions of KCN causes poisonous $HCN(g)$ to bubble out of the solution: $CN^-(aq) + H^+(aq) \rightarrow HCN(g)$.

4-6 (a)

$$P = \rho g h = (13.60 \times 10^3 \text{ kg m}^{-3})(9.807 \text{ m s}^{-2})(0.0950 \text{ m})$$

$$= 1.267 \times 10^4 \text{ Pa} = 0.125 \text{ atm}$$

(b) $\rho_1 g h_1 = \rho_2 g h_2$

$$h_2 = \frac{\rho_1}{\rho_2} h_1 = \left(\frac{13.60 \text{ g cm}^{-3}}{1.045 \text{ g cm}^{-3}} \right)(9.50 \text{ cm}) = 124 \text{ cm} = 1.24 \text{ m}$$

4-8 A 76-cm column of mercury exerts the same pressure at its base that the atmosphere does on the earth at sea level. The density of the mercury in such a column is 13.6 g cm^{-3} (which is the same as 13.6×10^3 g L^{-1}) and is uniform over the length of the column. If a fluid with a uniform density of only 1.3 g L^{-1} replaces the mercury, it clearly must be substantially longer to exert the same pressure. It is in fact longer in proportion to the ratio of the densities. The thickness of the hypothetical atmosphere is thus

$$76 \text{ cm} \times \left(\frac{13.6 \times 10^3}{1.3} \right) = 8.0 \times 10^5 \text{ cm} = 8.0 \text{ km}$$

Unlike the ocean (problem 4-7), the atmosphere is nowhere near uniform in density. Its attenuation with altitude means that it must be much thicker than 8 km to exert a pressure of 1 atm.

4-10 The conversions can be carried out using the information in Table 4-2 in the text:

$$5 \times 10^{-10} \text{ torr} \times \left(\frac{1 \text{ atm}}{760 \text{ torr}}\right) = 7 \times 10^{-13} \text{ atm}$$

$$5 \times 10^{-10} \text{ torr} \times \left(\frac{1 \text{ atm}}{760 \text{ torr}}\right) \times \left(\frac{1.01325 \times 10^5 \text{ Pa}}{1 \text{ atm}}\right) = 7 \times 10^{-8} \text{ Pa}$$

4-12 Boyle's law applies to an expansion at constant temperature. The computation is:

$$P_2 = \left(\frac{V_1}{V_2}\right) P_1 = \left(\frac{0.350 \text{ L}}{1.31 \text{ L}}\right) (1.23 \text{ atm}) = 0.329 \text{ atm}$$

4-14 Only in terms of the absolute temperature T is it correct to write

$$V_2 = \left(\frac{T_2}{T_1}\right) V_1$$

Converting the Celsius temperatures given in the problem to absolute temperatures and substituting gives

$$V_2 = \left(\frac{313.15 \text{ K}}{293.15 \text{ K}}\right) (4.00 \text{ L}) = 4.27 \text{ L}$$

4-16 The problem illustrates the operation of a gas thermometer. The temperature of a sample of an ideal gas is directly proportional to the absolute temperature as long as the pressure and the amount of gas in the sample do not change. Hence

$$T_2 = \left(\frac{V_2}{V_1}\right) T_1$$

If the temperature is converted from the Celsius scale to the Kelvin scale and the volumes are substituted:

$$T_2 = \left(\frac{5.26 \text{ L}}{5.40 \text{ L}}\right) \times 299.65 \text{ K} = 291.88 \text{ K}$$

Converting back to the Celsius scale gives 18.7°C.

4-18 This problem uses Charles's law, as in 4-16, but now a volume is to be determined:

$$V_2 = \left(\frac{T_2}{T_1}\right) V_1$$

T_2 is 503.15 K and T_1 is 293.15 K. Substituting gives

$$V_2 = \left(\frac{503.15 \text{ K}}{293.15 \text{ K}}\right) \times 3.41 \text{ L} = 5.85 \text{ L}$$

4-20 To prevent dangerous spurting from the container, the pressure inside must be brought below 0.96 atm. The pressure at 20°C (293.15 K) is 1.47 atm, and the pressure of an ideal gas at constant volume is directly proportional to the absolute temperature. Because $0.96/1.47 = 0.653$, the required absolute temperature is $0.653 \times 293.15 = 191.4$ K. Subtracting 273.15 K from this Kelvin temperature and multiplying the result by 1°C/1 K converts to -82°C.

4-22 The amount of gas n does not change when T and P change, so:

$$\frac{P_1 V_1}{T_1} = \frac{P_2 V_2}{T_2}$$

Solving for V_2 gives

$$V_2 = \left(\frac{P_1}{P_2}\right)\left(\frac{T_2}{T_1}\right) V_1$$

Substitution of $P_1 = 0.459$ atm, $P_2 = 0.980$ atm, $T_1 = 573.15$ K, $T_2 = 673.15$ K, and $V_1 = 63.6$L gives $V_2 = 35.0$L.

4-24 (a) First we calculate the chemical amount of O_2 corresponding to 0.30 kg of O_2. It is 9.375 mol O_2. Assuming that O_2 is an ideal gas, we then calculate its pressure in the scuba tank:

$$\begin{aligned}
P &= \frac{nRT}{V} \\
&= \frac{9.375 \text{ mol } (0.08206 \text{ L atm mol}^{-1}\text{K}^{-1})278.15 \text{ K}}{2.32 \text{ L}} = 92 \text{ atm} \\
&= 1.4 \times 10^3 \text{ psi}
\end{aligned}$$

(b)

$$V_2 = \left(\frac{P_1}{P_2}\right)\left(\frac{T_2}{T_1}\right) V_1 = \left(\frac{92 \text{ atm}}{0.98 \text{ atm}}\right)\left(\frac{303 \text{ K}}{278 \text{ K}}\right)(2.32 \text{ L}) = 2.4 \times 10^2 \text{ L}$$

4-26 (a) The chemical equation is $2\,Al + 6\,HCl \rightarrow 3\,H_2 + 2\,AlCl_3$.
(b) Assume that the hydrogen behaves ideally and substitute the volume, pressure, and absolute temperature into the ideal gas equation:

$$n = \frac{PV}{RT} = \frac{(0.750 \text{ L})(10.0 \text{ atm})}{(0.08206 \text{ L atm mol}^{-1}\text{K}^{-1})(303.15 \text{ K})} = 0.3015 \text{ mol}$$

According to the balanced equation, 2/3 this amount of aluminum must react. The chemical amount of Al is therefore 0.2010 mol, and the mass of Al is 0.2010 mol multiplied by 26.98 g mol^{-1}. This answer is 5.42 g Al.

4-28 (a) $Fe(s) + H_2SO_4(aq) \rightarrow H_2(g) + FeSO_4(aq)$.

(b) The 300×10^3 g mass of sulfuric acid is converted to chemical amount by dividing by its molar mass (98.08 g mol^{-1}). The equation states that 1 mol of H_2 forms for every 1 mol of H_2SO_4, so the reaction gives 3.059×10^3 mol of H_2. Substituting this value as n in the ideal-gas equation with $T = 300$ K and $P = 1.0$ atm gives $V = 7.5 \times 10^4$ L.

(c) The formula $V = 4/3\pi r^3$ relates the volume of a sphere to its radius. A liter is equal to a cubic decimeter, so inserting the volume of gas in liters into this formula and solving for r gives the radius of the spherical balloon in decimeters. It is 26 dm, which is 2.6 m (8.6 feet).

4-30 Using the ideal gas law for chlorine gives its number of moles as

$$n = \frac{PV}{RT} = \frac{(0.953 \text{ atm})(5.32 \text{ L})}{(0.08206 \text{ L atm mol}^{-1}\text{K}^{-1})(306.15 \text{ K})} = 0.2018 \text{ mol}$$

This is also the chemical amount of MnO_2 reacting, because one mole of MnO_2 generates one mole of Cl_2. The mass is then found by multiplying by the molar mass of MnO_2, 86.937 g mol^{-1}, to give 17.5 g MnO_2.

4-32 Divide the mass of KO_3, 4.69 g, by its molar mass, 87.10 g mol^{-1}, to find the chemical amount of KO_3, 0.05385 mol. The chemical amount of ozone is then

$$\left(\frac{5 \text{ mol } O_3}{2 \text{ mol } KO_3} \right) \times (0.05385 \text{ mol } KO_3) = 0.1346 \text{ mol } O_3$$

From the ideal gas law,

$$V = \frac{nRT}{P} = \frac{(0.1346 \text{ mol})(0.08206 \text{ L atm mol}^{-1}\text{K}^{-1})(258.15 \text{ K})}{0.134 \text{ atm}} = 21.3 \text{ L}$$

4-34 The total chemical amount of gas in the reaction mixture is $13 + 31 + 93 = 137$ mol. The ammonia contributes 13 mol so its mole fraction is $13/137 = 0.095$. The partial pressure of the ammonia is simply its mole fraction multiplied by the total pressure, as long as the mixture behaves according to Dalton's law. Then $P_{NH_3} = 20$ atm.

4-36 For ideal gas mixtures, mole fractions are the same as fractions by volume, so the mole fraction of N_2 is 0.035. Its partial pressure is the mole fraction multiplied by the total pressure, giving 3.2 atm.

4-38 (a) Before the reaction, the mole fraction of the Br_2 is $4.5/(4.5 + 33.1) = 0.12$.

(b) According to the balanced equation, the formation of 2.2 mol of BrF_5 consumes 1.1 mol of Br_2 and 5.5 mol of F_2. At the indicated point in the reaction, there are 3.4 mol of Br_2 and 27.6 mol of F_2 left. This plus the 2.2 mol of BrF_5 means that there is 33.2 mol of substances of all kinds present. The mole fraction of Br_2 is $3.4/33.2 = 0.10$. Despite the fact that about a quarter of the Br_2 has been consumed, the mole fraction of Br_2 has dropped by only a sixth.

4-40 (a)

$$n = \frac{PV}{RT} = \frac{(0.200 \text{ atm})(1.500 \text{ L})}{(0.08206 \text{ L atm mol}^{-1}\text{K}^{-1})(313.15 \text{ K})} = 1.167 \times 10^{-2} \text{ mol } O_2$$

$$N = (1.167 \times 10^{-2} \text{ mol})(6.022 \times 10^{23} \text{ mol}^{-1}) = 7.03 \times 10^{21} \text{ molecules}$$

(b) Oxygen is the limiting reactant.

$2 H_2(g) + O_2(g) \rightarrow 2 H_2O(l)$

Moles $H_2O = \left(\frac{2 \text{ mol } H_2O}{1 \text{ mol } O_2}\right)(1.167 \times 10^{-2} \text{ mol } O_2) = 2.335 \times 10^{-2} \text{ mol}$

$(2.335 \times 10^{-2} \text{ mol})(18.02 \text{ g mol}^{-1}) = 0.421 \text{ g water}$

4-42 The relationship between the root-mean-square speed of the molecules of a gas and the temperature of the gas is

$$u_{rms} = \sqrt{\frac{3RT}{\mathcal{M}}}$$

where $\mathcal{M}$ is the molar mass of the gas. We substitute $\mathcal{M} = 0.023 \text{ kg mol}^{-1}$, $R = 8.315 \text{ J mol}^{-1} \text{ K}^{-1}$, and $T = 0.00024 \text{ K}$ to compute $u_{rms} = 0.51 \text{ m s}^{-1}$. Common errors here are to use $\mathcal{M}$ in g mol^{-1} instead of kg mol^{-1} or to use R in the wrong units.

4-44 To calculate the rms speeds of helium, argon, and xenon atoms at 2000 K we substitute the molar masses in kg mol^{-1} of the three gases successively into the expression

$$u_{rms} = \sqrt{\frac{3RT}{\mathcal{M}}}$$

taking $T = 2000 \text{ K}$ and $R = 8.315 \text{ J mol}^{-1} \text{ K}^{-1}$. The answers are 3.53 km s^{-1} for helium, 1.12 km s^{-1} for argon and 0.616 km s^{-1} for xenon. These values are respectively 31.5%, 10.0% and 5.50% of the earth's escape velocity. Helium is much more likely to escape than the heavier gases.

4-46 Molecules of Cl_2 have a greater mass than those of ClO_2. At a given temperature, the speed distribution of heavier molecules is shifted to lower values. Thus the percentage of chlorine molecules with speeds in excess of 400 m s^{-1} will be less than 35%.

4-48   Moles escaping $= \dfrac{(0.001 \text{ atm})(0.200 \text{ L})}{(0.08206 \text{ L atm mol}^{-1}\text{K}^{-1})(298 \text{ K})} = 8.18 \times 10^{-6}$ mol

$$Z_w = \text{molecules escaping per second}$$
$$= \frac{(8.18 \times 10^{-6} \text{ mol})(6.022 \times 10^{23} \text{ mol}^{-1})}{3600 \text{ s}} = 1.37 \times 10^{15} \text{ s}^{-1}$$

$$= \tfrac{1}{4}\frac{N}{V}\bar{u}A = \tfrac{1}{4}\frac{N_0 P}{RT}\bar{u}A$$

$$\bar{u} = \sqrt{\frac{8RT}{\pi\mathcal{M}}} = \sqrt{\frac{8(8.3145 \text{ J mol}^{-1}\text{ K}^{-1})(298 \text{ K})}{\pi(2.016 \times 10^{-3} \text{ kg mol}^{-1})}} = 1770 \text{ m s}^{-1}$$

$$A = \frac{4RT}{N_0 P} \frac{Z_w}{\bar{u}}$$
$$= \frac{4(8.3145 \text{ J mol}^{-1}\text{ K}^{-1})(298 \text{ K})(1.37 \times 10^{15} \text{ s}^{-1})}{(6.022 \times 10^{23} \text{ mol}^{-1})(0.990 \text{ atm})(101325 \text{ Pa atm}^{-1})(1770 \text{ m s}^{-1})}$$
$$= 1.27 \times 10^{-13} \text{ m}$$

Radius $= \sqrt{\frac{A}{\pi}} = 2 \times 10^{-7}$ m

The proper choice of units for R is crucial in this problem.

4-50 The problem is a straightforward application of Graham's law of effusion. The relative rates of effusion are given by

$$\frac{\text{rate}_{F_2}}{\text{rate}_{BrF_5}} = \sqrt{\frac{\mathcal{M}_{BrF_5}}{\mathcal{M}_{F_2}}}$$

Substituting the molar masses gives 2.145 as the ratio of the rates of effusion. As effusion continues the remaining mixture becomes enriched in the heavier gas, and this ratio drops.

4-52

$$\frac{\text{rate (He)}}{\text{rate (H}_2)} = 3 = \frac{N(\text{He})}{N(\text{H}_2)}\sqrt{\frac{m(\text{H}_2)}{m(\text{He})}}$$

$$\frac{N(\text{He})}{N(\text{H}_2)} = 3\sqrt{\frac{m(\text{He})}{m(\text{H}_2)}} = 3\sqrt{1.986} = 4.227$$

$$X(\mathrm{H_2}) = \frac{N(\mathrm{H_2})}{N(\mathrm{H_2}) + N(\mathrm{He})} = \frac{1}{1 + 4.23} = 0.191$$

The enrichment factor per stage is $\sqrt{1.986} = 1.409$. After n stages, the ratio is

$$\left(\frac{N(\mathrm{H_2})}{N(\mathrm{He})}\right)_n = \left(\frac{N(\mathrm{H_2})}{N(\mathrm{He})}\right)_0 (1.409)^n = \frac{1}{4.227}(1.409)^n$$

For 99.9% purity,

$$\frac{N(\mathrm{H_2})}{N(\mathrm{He})} = \frac{99.9}{0.1} = 999$$

$$999 = \frac{1}{4.227}(1.409)^n$$

$$4223 = (1.403)^n$$

$$\log_{10} 4223 = n \log_{10} 1.403$$

$$n = 24.7 \quad \text{Thus 25 stages are required.}$$

4-54

$$\lambda = \frac{1}{\sqrt{2}\pi d^2 N/V} = \frac{1}{\sqrt{2}\pi d^2 N_0 P/RT}$$

$$P = \frac{RT}{\sqrt{2}\pi d^2 \lambda N_0} = \frac{(8.315 \text{ J mol}^{-1} \text{ K}^{-1})(300 \text{ K})}{\sqrt{2}\pi (3.16 \times 10^{-10} \text{ m})^3 (6.022 \times 10^{23} \text{ mol}^{-1})}$$

$$= 2.95 \times 10^7 \text{ Pa} = 292 \text{ atm}$$

At this temperature for Kr, $\bar{u} = \sqrt{8RT/\pi\mathcal{M}} = 275.3 \text{ m s}^{-1}$

$$D = \frac{3\pi}{16}\lambda\bar{u} = 5.1 \times 10^{-8} \text{ m}^2 \text{ s}^{-1}$$

4-56 We solve the van der Waals equation for the pressure:

$$P = \frac{nRT}{V - nb} - a\left(\frac{n^2}{V^2}\right)$$

Substituting the tabulated values of a and b together with $V = 2\,500 \times 10^3$ L, $R = 0.08206$ L atm mol^{-1}K^{-1}, $n = 140 \times 10^6/18.0153$ mol, and $T = 813.15$ K gives $P = 176$ atm, or 2590 psi.

4-58 (a) $\dfrac{60.0 \text{ g}}{16.04 \text{ g mol}^{-1}} = 3.74 \text{ mol}$

$$T = \frac{PV}{nR} = \frac{(130 \text{ atm})(1.00 \text{ L})}{(3.74 \text{ mol})(0.08206 \text{ L atm mol}^{-1}\text{K}^{-1})} = 424 \text{ K}$$

(b) $(P + an^2/V^2)(V - nb) = nRT$

$$T = \frac{(130 + 2.253(3.74)^2)(1.00 - (3.74)(0.04278))}{(3.74)(0.08206)} = 442 \text{ K}$$

Because 442 K > 424 K, the a term (attractive forces) dominates under these conditions.

4-60

$$\begin{aligned}
P &= \rho g h \\
&= (4.9 \times 10^3 \text{ kg m}^{-3})(9.80665 \text{ m s}^{-2})(9.0 \text{ ft})(0.3048 \text{ m ft}^{-1}) \\
&= 1.32 \times 10^5 \text{ kg m}^{-1} \text{ s}^{-2} = 1.32 \times 10^5 \text{ Pa} = 1.3 \text{ atm}
\end{aligned}$$

$$(1.3 \text{ atm})(14.696 \text{ lb in}^{-2} \text{ atm}^{-1}) = 19 \text{ psi}$$

4-62 Setting $V = 0$ gives

$$\begin{aligned}
-209.4 \text{ L} &= 0.456 \text{ L } {}^{\circ}\text{F}^{-1} \, t_f \\
t_f &= -459 {}^{\circ}\text{F}
\end{aligned}$$

4-64 Density $\propto \dfrac{n}{V} = \dfrac{P}{RT}$

At constant P, T changes from 323 K to 423 K. The density will be multiplied by the ratio $\frac{323}{423}$.

$$\text{Density} = \left(\frac{323 \text{ K}}{423 \text{ K}}\right)(2.94 \text{ g L}^{-1}) = 2.24 \text{ g L}^{-1}$$

Consider one liter of gas at 50°, weighing 2.94 g. Then

$$n = \frac{PV}{RT} = \frac{(1.00 \text{ atm})(1.00 \text{ L})}{(0.08206 \text{ L atm}^{-1} \text{ mol}^{-1} \text{ K}^{-1})(323 \text{ K})}$$

$$= 3.77 \times 10^{-2} \text{ mol}$$

$$\text{molar mass} = \frac{2.94 \text{ g}}{3.77 \times 10^{-2} \text{ mol}} = 77.9 \text{ g mol}^{-1}$$

4-66 (a) Assume that the dry air is an ideal gas. Its density is determined by combining the ideal gas equation with the relation $n = m/M$ (where m is the mass of the gas, and M is the molar mass of the gas) and the definition of density ($\rho = m/V$) and rearranging to give

$$\rho = \frac{MP}{RT}$$

A temperature of 95°F is 35.0°C or 308.15 K. Substituting this temperature, $P = 1.00$ atm, $M = 29.0$ g mol^{-1}, and $R = 0.08206$ L atm mol^{-1}K^{-1} gives $\rho = 1.15$ g L^{-1}.

(b) A temperature of 50°F is 10.0°C or 283.15 K. The method of part (a) gives the density of the cool dry air as $\rho = 1.25$ g L^{-1}.

(c) Saturating dry air with water vapor means adding a component to the mixture that has a molar mass less that the M's of N_2 and O_2, the main components of dry air. The effective molar mass of moist air is therefore less than 29.0 g mol^{-1}. The density of a gas is directly proportional to its molar mass M (see part a). It follows that moist air is less dense than dry air at any given T and P. If a batted ball truly carries better in low-density air, then high humidity favors the home run.

4-68 A cubic foot is 28.317 L according to tables of conversion factors. This factor can also be computed: there are 12^3 cubic inches in a cubic foot and 2.54^3 cm^3 in a cubic inch, so a cubic foot is $12^3 \times 2.54^3 = 28\,316.8$ cm^3 or 28.317 L. The 1.0 lb of Hydrone generates 2.6×28.317 L of hydrogen. Solving the ideal gas law for the chemical amount gives 3.285 mol of H_2. The balanced equation states that it takes 2 mol of Na to generate 1 mol of H_2. Hence, the Hydrone contains 6.57 mol of sodium. Because the molar mass of sodium is 23 g mol^{-1}, the 1.0 lb (453.59 g) of Hydrone contained 151 g of sodium. This is 33 % of the mass of the Hydrone.

4-70 For every six molecules originally present, three (half) are consumed. These three react to give two new ones for a total of five molecules. If n is multiplied by $\frac{5}{6}$ with V and T constant, then P must be multiplied by $\frac{5}{6}$. This gives

$$P = \frac{5}{6}(0.740 \text{ atm}) = 0.617 \text{ atm}$$

4-72 $CS_2(g) + 3\,O_2(g) \rightarrow CO_2(g) + 2\,SO_2(g)$

For every mole of $CS_2(g)$ that reacts, the total number of moles changes by one. Because the temperature and volume are the same before and after, the initial partial pressure of CS_2 must be equal in magnitude to the change in total pressure, namely 3.00 - 2.40 = 0.60 atm.

$$n = \frac{PV}{RT} = \frac{(0.60 \text{ atm})(10.0 \text{ L})}{(0.08206 \text{ L atm mol}^{-1}\text{K}^{-1})(373.15 \text{ K})} = 0.196 \text{ mol}$$

$$\text{mass} = (0.196 \text{ mol})(76.14 \text{ g mol}^{-1}) = 14.9 \text{ g}$$

4-74 (a) The volume of the trap is expressed in liters ($1 \text{ L} = 10^{-3} \text{ m}^3$) and the amount of its contents is expressed in moles ($n = 500/6.022 \times 10^{23} \text{ mol}^{-1} = 8.3029 \times 10^{-22}$). Then, the problem becomes a substitution into $PV = nRT$. Completing the arithmetic gives $P = 1.6 \times 10^{-14}$ atm. The pressure is very low. Otherwise the frail walls made of light would not contain the gas.

(b) The mean free path of the sodium atoms in the optical trap is

$$\lambda = \frac{RT}{\sqrt{2}\pi d^2 N_o P} = 3.9 \text{ m}$$

Solving for the diameter d gives 3.4×10^{-10} m. At room conditions in gaseous sodium the mean free path is about 8×10^{-8} m, smaller by a factor of 50,000,000.

4-76 (a) kinetic energy per mole $= \frac{3}{2}RT$

average kinetic energy per molecule $= \frac{3}{2}\frac{RT}{N_0}$

$$= \frac{3}{2}\frac{(8.3145 \text{ J mol}^{-1} \text{ K}^{-1})(473 \text{ K})}{6.022 \times 10^{23} \text{ mol}^{-1}} = 9.80 \times 10^{-21} \text{ J at 200°C}$$

At 400°C, this gives 1.39×10^{-20} J. Both results are independent of mass, and thus of oxygen isotopic composition.

(b) The average speed is

$$\bar{u} = \sqrt{\frac{8RT}{\pi \mathcal{M}}}$$

At 200°C, the combination of constants is

$$\frac{8RT}{\pi} = \frac{8(8.3145 \text{ J mol}^{-1} \text{ K}^{-1})(473 \text{ K})}{\pi} = 1.001 \times 10^4 \text{ J mol}^{-1}$$

The molar mass of the lightest species, $^{16}O^{16}O$, is (approximately) 32.0 g mol^{-1} and that of the heaviest, $^{18}O^{18}O$, is 36.0 g mol^{-1}. This gives average speeds of

$$\bar{u} = \sqrt{\frac{1.001 \times 10^4 \text{ J mol}^{-1}}{32.0 \times 10^{-3} \text{ kg mol}^{-1}}} = 559 \text{ m s}^{-1} \text{ for the lightest,}$$

and 527 m s^{-1} for the heaviest. At 400°C, the corresponding average speeds are 667 and 629 m s^{-1}.

4-78 (a) From the right angle geometry in the figure,

$$r \cos \theta = \Delta l / 2$$
$$\Delta l = 2r \cos \theta$$

(b) Magnitude of momentum $= mu$

Component perpendicular to wall $= mu \cos \theta$

$$\Delta p_{\text{wall}} = mu \cos \theta - (-mu \cos \theta) = 2mu \cos \theta$$

(c) Time between collisions $= \dfrac{\Delta l}{u} = \Delta t$

Force on wall $= \dfrac{\Delta p_{\text{wall}}}{\Delta t} = \dfrac{2mu \cos \theta}{2r \cos \theta / u} = \dfrac{mu^2}{r}$

(d) From N molecules,

$$\text{force} = \frac{N m \overline{u^2}}{r}$$

The surface area of the sphere is $4\pi r^2$, so

$$P = \frac{\text{force}}{\text{area}} = \frac{N m \overline{u^2}}{4\pi r^2 (r)}$$

The volume of the sphere is $V = \frac{4}{3}\pi r^3$. Multiplying by P gives

$$PV = \frac{1}{3} N m \overline{u^2}$$

4-80 The collision rate with the wall (in molecules per second) is

$$Z_{\text{W}} = \frac{1}{4} \frac{N}{V} \sqrt{\frac{8RT}{\pi \mathcal{M}}} A$$

In moles per second it is (from the ideal gas law)

$$\frac{1}{4} \frac{n}{V} \sqrt{\frac{8RT}{\pi \mathcal{M}}} A = \frac{1}{4} P \sqrt{\frac{8}{\pi \mathcal{M} RT}} A$$

$$P = 2000 \text{ psi} \left(\frac{1 \text{ atm}}{14.696 \text{ psi}} \right) \left(\frac{101325 \text{ Pa}}{1 \text{ atm}} \right) = 1.38 \times 10^7 \text{ Pa}$$

$$\mathcal{M} = 16.04 \times 10^{-3} \text{ kg mol}^{-1}$$

$$A = 1.0 \text{ mm}^2 = 1.0 \times 10^{-6} \text{ m}^2$$

$$\text{Rate} = \frac{1}{4}(1.38 \times 10^7 \text{ Pa})\sqrt{\frac{8}{\pi(16.04 \times 10^{-3} \text{ kg mol}^{-1})(8.3145)(293 \text{ K})}}(1 \times 10^{-6} \text{ m}^2)$$

$$= 0.88 \text{ mol s}^{-1}$$

$$(0.88 \text{ mol s}^{-1})(60 \times 60 \times 24 \text{ s day}^{-1}) = 7.6 \times 10^4 \text{ mol day}^{-1}$$

$$(7.6 \times 10^4 \text{ mol day}^{-1})(16.04 \text{ g mol}^{-1}) = 1.2 \times 10^6 \text{ g day}^{-1}$$

$$\text{Volume of tank} = \pi(20)^2(50) \text{ ft}^3 \times (0.3048 \text{ m ft}^{-1})^3$$
$$= 1.78 \times 10^3 \text{ m}^3 = 1.78 \times 10^6 \text{ L}$$

$$n = \frac{PV}{RT} = \frac{(2000 \text{ psi}/14.696 \text{ psi atm}^{-1})(1.78 \times 10^6 \text{ L})}{(.08206 \text{ L atm mol}^{-1}\text{K}^{-1})(293 \text{ K})} = 1.01 \times 10^7 \text{ mol}$$

$$\text{Fraction lost per day} = \frac{7.6 \times 10^4 \text{ mol}}{1.01 \times 10^7 \text{ mol}} = 0.0075$$

4-82 (a) $\dfrac{N}{V} = \dfrac{N_0 n}{V} = N_0 \dfrac{P}{RT}$

$$D = \frac{3}{8}\sqrt{\frac{RT}{\pi\mathcal{M}}}\frac{1}{d^2 N_0 \frac{P}{RT}} = \frac{3}{8}\frac{(RT)^{\frac{3}{2}}}{(\pi\mathcal{M})^{\frac{1}{2}}d^2 N_0 P}$$

Expressing everything in SI units gives

$$D = \frac{3}{8}\frac{[(8.315 \text{ J mol}^{-1}\text{ K}^{-1})(293 \text{ K})]^{\frac{3}{2}}}{[(\pi)(17 \times 10^{-3} \text{ kg mol}^{-1})]^{\frac{1}{2}}(3 \times 10^{-10} \text{ m})^2(6.02 \times 10^{23} \text{ mol}^{-1})(101325 \text{ Pa})}$$

$$= 3.55 \times 10^{-5} \text{ m}^2 \text{ s}^{-1}$$

(b)

$$t = \frac{\overline{\Delta r^2}}{6D} = \frac{(100 \text{ m})^2}{6(3.55 \times 10^{-5} \text{ m}^2 \text{ s}^{-1})} = 4.7 \times 10^7 \text{ s} = 1.5 \text{ years}$$

4-84 (a) It is easiest to convert the number density to density in moles per liter:

$$\frac{n}{V} = \frac{10 \text{ atoms}}{1 \text{ cm}^3} \times \frac{1 \text{ mol}}{6.022 \times 10^{23} \text{ atoms}} \times \frac{1000 \text{ cm}^3}{1 \text{ L}} = 1.66 \times 10^{-20} \text{ mol L}^{-1}$$

Substitution of this value into $P = (n/V)RT$ gives $P = 1.4 \times 10^{-19}$ atm.

(b) The root-mean-square speed is calculated by substituting $T = 100$ K and $\mathcal{M} = 1.008 \times 10^{-3}$ kg mol^{-1} into

$$u_{rms} = \sqrt{\frac{3RT}{\mathcal{M}}}$$

being sure to use R in the proper units (J mol^{-1} K^{-1}). The answer is 1.57×10^3 m s^{-1}. Multiplying this speed by the time between collisions gives the distance between collisions, on the average. It is 1.57×10^{12} m, about 10.5 times the distance between the earth and the sun.

4-86 From Eq. 4-15, the rate at which a molecule collides with other molecules is

$$Z_1 = 4\frac{N}{V}d^2\sqrt{\frac{\pi RT}{\mathcal{M}}} = 4N_0\frac{P}{RT}d^2\sqrt{\frac{\pi RT}{\mathcal{M}}}$$

(a) Inserting $d = \sigma = 3.82 \times 10^{-10}$ m for methane, at 25°C and 1 atm pressure (101,325 Pa), gives a rate of 1.0×10^{10} s^{-1}.

(b) At $P = 1.0 \times 10^{-7}$ atm, the rate is 1.0×10^3 s^{-1}, smaller by a factor of 10^7

4-88

$$V_{\text{LJ}} = 4\epsilon\left[\left(\frac{\sigma}{R}\right)^{12} - \left(\frac{\sigma}{R}\right)^{6}\right]$$

$$F(R) = -\frac{dV(R)}{dR} = \frac{4\epsilon}{\sigma}\left[12(\sigma/R)^{13} - 6(\sigma/R)^{7}\right]$$

$$\sigma = 3.40\,\text{Å} - 3.40 \times 10^{-10} \text{ m}$$

$$\epsilon = 1.654 \times 10^{-21} \text{ J}$$

$$F(R) = \frac{24\epsilon}{\sigma}\left[2(\sigma/R)^{13} - (\sigma/R)^{7}\right] = (1.168 \times 10^{-10} \text{ J m}^{-1})\left[2(\sigma/R)^{13} - (\sigma/R)^{7}\right]$$

At $3.0\,\text{Å} = 3.0 \times 10^{-10}$ m, $F(R) = 9.08 \times 10^{-10}$ J m^{-1}
$3.4\,\text{Å} = 3.4 \times 10^{-10}$ m, $F(R) = 1.17 \times 10^{-10}$ J m^{-1}
$3.8\,\text{Å} = 3.8 \times 10^{-10}$ m, $F(R) = 1.40 \times 10^{-12}$ J m^{-1}
$4.2\,\text{Å} = 4.2 \times 10^{-10}$ m, $F(R) = -1.16 \times 10^{-11}$ J m^{-1}

The force is repulsive at the first three distances, attractive at the fourth.

4-90

$$\frac{\text{Volume of chlorine}}{\text{Volume of compound}} = \frac{0.688 \text{ L}}{0.153 \text{ L}} = \frac{\text{Moles of Cl}_2}{\text{Moles of compound}} = 4.5$$

Each mole of compound thus contains $2 \times 4.5 = 9$ mol of Cl atoms. The chemical amount of compound comes from the ideal gas law:

$$n = \frac{PV}{RT} = \frac{(1.00 \text{ atm})(0.153 \text{ L})}{(0.08206 \text{ L atm mol}^{-1}\text{K}^{-1})(273.15 \text{ K})} = 6.83 \times 10^{-3} \text{ mol}$$

The molar mass is the mass (2.842 g) divided by the chemical amount (6.83×10^{-3} mol), giving 416 g mol^{-1}. Subtracting 9×35.453 g mol^{-1} from the Cl leaves 97.3 g mol^{-1}. Dividing by the molar mass of B, 10.811 g mol^{-1}, gives 9.0, so the formula is B_9Cl_9. Note the similarities and differences between this problem and problem 2-44.

4-92 (a) The reaction is

$$Rb_2SO_3 + 2 \text{ HBr} \rightarrow 2 \text{ RbBr} + H_2O + SO_2$$

The initial chemical amount of Rb_2SO_3 is

$$\frac{6.24 \text{ g}}{251.0 \text{ g mol}^{-1}} = 0.02486 \text{ mol } Rb_2SO_3$$

The initial chemical amount of HBr is

$$n = \frac{PV}{RT} = \frac{(0.953 \text{ atm})(1.38 \text{ L})}{(0.08206 \text{ L atm mol}^{-1}\text{K}^{-1})(348.15 \text{ K})} = 0.04603 \text{ mol HBr}$$

Because 2 mol of HBr reacts with 1 mol of Rb_2SO_3, the HBr will be used up first. It is the limiting reactant.

(b) Each mole of HBr that reacts should theoretically give one mole of RbBr. The theoretical yield is

$$(0.04603 \text{ mol RbBr})(165.37 \text{ g mol}^{-1}) = 7.61 \text{ g}$$

(c) The percentage yield is $\frac{7.32 \text{ g}}{7.61 \text{ g}} \times 100\% = 96.2\%$

Chapter 5

Solids, Liquids, and Phase Transitions

5-2 Because the substance is viscous and nearly incompressible, it is not a gas. Elasticity is closely related to rigidity and hardness, properties that are characteristic of solids rather than liquids. Hence, the substance is a liquid.

5-4 (a) The density of this substance is $57.9/18.3 = 3.16$ g $L^{-1} = 0.00316$ g cm^{-3}. Although the conditions of temperature and pressure are not given, it is hard to imagine a solid or liquid having such a low density regardless of conditions. Thus, a key to the problem is a knowledge of typical densities such as those of ice and water (approximately 0.9 and 1.0 g cm^{-3}). Demonstrations are very useful in teaching this.

(b) The molar volume is the volume occupied by one mole of a substance. It equals the molar mass divided by the density or, equivalently, the molar mass multiplied by the reciprocal of the density:

$$\left(\frac{123 \text{ g}}{1 \text{ mol}}\right) \times \left(\frac{18.3 \times 10^3 \text{ cm}^3}{57.9 \text{ g}}\right) = 3.89 \times 10^4 \frac{\text{cm}^3}{\text{mol}}$$

This large molar volume is quite consistent with the value quoted in the text for gases under typical conditions.

5-6 The same cooling of an ideal gas from 343.15 K to 283.15 K would decrease its volume to 0.825 of its original volume, where 0.825 is obtained as the ratio of 343.15 K to 283.15 K. This substance in fact decreases its volume to 0.816 of its original value. It is a nearly ideal gas.

5-8 In a gas there are only weak or nearly nonexistent intermolecular attractions so the molecules are not held strongly to each other. In solid and liquid metals, the intermolecular forces are large, so that the metal changes its volume little even with considerable increase in the temperature.

5-10 The surface tension of liquid NaCl should be higher than that of CCl_4 because the intermolecular forces in NaCl (ion-ion forces) are intrinsically stronger than the forces in CCl_4 (dispersion forces).

5-12 The kinetic theory predicts an increase in molecular speed in all three states of matter with increasing temperature. Since diffusion depends on the random motions of the molecule of substances, which become more rapid at higher temperature, the theory predicts that the diffusion constants should increase with temperature in all states of matter.

5-14 Dipole-dipole forces arise among particles with permanent dipoles, that is, among particles having pre-existent nonuniform charge distributions. Dispersion forces are sometimes called instantaneous dipole-induced dipole forces. They operate at shorter range than dipole-dipole forces and are weaker. Dipole-dipole forces are found in polar substances (such as $HCl(g)$); dispersion forces are found in all substances.

5-16 (a) dispersion forces   (b) dipole-dipole forces (predominant) and dispersion forces   (c) dispersion forces   (d) ion-ion forces (predominant) and dispersion forces

5-18 An atom of argon should be most strongly attracted by an atom of krypton. The krypton atom has more electrons and is more polarizable than one of argon and the strength of dispersion forces depends on the polarizability of the interacting species.

5-20 False. A small atom such as hydrogen can approach another atom quite closely, whereas a weak chemical bond between a pair of large atoms can leave the atoms quite far apart.

5-22 The pentafluoride with the largest molar mass (IF_5) is the solid, the one with the second largest molar mass (BrF_5) is the liquid, and the one with the least molar mass (ClF_5) is the gas; the strength of attractive forces increases with increasing molar mass.

5-24 $He < Ar < SO_2 < HF < CaF_2$

Ar is heavier than He, while SO_2 is non-spherical and interacts through electrostatic forces. Hydrogen bonding is present for HF, while CaF_2 is bound by ionic forces.

5-26 The structure has chains of H—O—F molecules, which are bent, linked by O···H hydrogen bonds. The fluorine atoms stick out to the side of the chain.

5-28 Hydrogen peroxide will have extensive hydrogen bonding and display anomalies similar to those displayed by water. It should have a higher boiling point than F_2 and H_2S.

5-30 In HF(l), one hydrogen bond is possible for every molecule of HF; in $H_2O(l)$, two hydrogen bonds are possible for every molecule of H_2O. The answers are therefore N_0 and $2N_0$.

5-32 Assume that helium vapor in equilibrium with liquid helium can be described by the ideal-gas equation. Rearranging the equation so that the molar volume is explicitly on the left and substituting the values from the problem gives

$$\frac{V}{n} = \frac{RT}{P} = \frac{(0.08206 \text{ L atm mol}^{-1}\text{K}^{-1})2.20 \text{ K}}{0.05256 \text{ atm}} = 3.43 \text{ L mol}^{-1}$$

Even at these extreme conditions, the molar volume is still large, a characteristic of a gas. It is, however, smaller than that at STP, 22.4 L mol^{-1}.

5-34 The tungsten vapor can be treated as an ideal gas. The molar volume equals to volume of a sample divided by the number of moles present. Then:

$$\frac{V}{n} = \frac{RT}{P} = \frac{(0.08206 \text{ L atm mol}^{-1}\text{K}^{-1})2773 \text{ K}}{7.0 \times 10^{-9} \text{ atm}} = 3.25 \times 10^{10} \text{ L mol}^{-1}$$

If the vapors occupy 3.25×10^{10} L per mole they contain 1 mol per 3.25×10^{10} L. But a mole of tungsten is 6.022×10^{23} atoms so there are 1.9×10^{13} atoms per liter, or 1.9×10^{10} per cm^3.

5-36 The partial pressure of the hydrogen is the total pressure minus the vapor pressure of water at this temperature, that is:

$$P_{H_2} = P_{\text{tot}} - P_{H_2O} = 0.9900 - 0.0313 = 0.9587 \text{ atm}$$

The chemical amount of hydrogen in a 1.000 L sample of the wet hydrogen is

$$n_{H_2} = \frac{PV}{RT} = \frac{(0.9587 \text{ atm})(1.000 \text{ L})}{(0.08206 \text{ L atm mol}^{-1}\text{K}^{-1})(298.15 \text{ K})} = 0.039185 \text{ mol } H_2$$

The mass of hydrogen is this chemical amount multiplied by 2.01588 g mol^{-1}, the molar mass of H_2. The answer is 0.07899 g of H_2.

5-38 The chemical amount of the ammonia that is produced can be determined from the stoichiometry of the balanced equation

$$3.68 \text{ g NH}_4\text{Cl} \times \frac{1 \text{ mol NH}_4\text{Cl}}{53.491 \text{ g NH}_4\text{Cl}} \times \frac{1 \text{ mol NH}_3}{1 \text{ g NH}_4\text{Cl}} = 0.06880 \text{ mol NH}_3$$

This amount of ammonia exerts a partial pressure of 0.9465 atm in the mixture; the water vapor exerts the other 0.0419 atm. According to Dalton's law the water vapor and the ammonia do not interact with each other in the container but instead effectively interpenetrate. The volume of the wet ammonia therefore equals the volume that dry ammonia would occupy if present by itself. This volume is calculated using the chemical amount of ammonia and the partial pressure of ammonia in the ideal gas equation:

$$V = \frac{nRT}{P} = \frac{(0.06880 \text{ mol})(0.08206 \text{ L atm mol}^{-1}\text{K}^{-1})(303.15 \text{ K})}{(0.9465 \text{ atm})} = 1.81 \text{ L}$$

5-40 Reading from the figure across on the $P = 4$ atm line to the liquid/vapor equilibrium line and then down to the temperature axis gives an estimate of 430 K or 157°C for the boiling point of water in the pressure cooker.

5-42 The interatomic forces in aluminum are stronger because it melts and boils higher than thallium. Hence, we expect the vapor pressure of thallium to be higher than that of aluminum at room temperature.

5-44 Gray tin is favored over white tin by lower temperature, but white tin is favored by higher pressure (because it is more dense than gray tin). Suppose the two forms of tin are present at equilibrium at 1 atm and 13.2°C. Raising the pressure to 2 atm (eventually) converts all the tin to white tin. To restore the gray tin the temperature must be adjusted in the direction that favors gray tin, that is, the temperature must be lowered below 13.2°C.

5-46 The triple point and critical point are joined by a line on the PT graph that represents the conditions at which liquid and gaseous N_2 are in equilibrium. The normal boiling point is on this line. The line curves, but its curvature cannot be determined from the available data. The solid/liquid equilibrium line extends from the triple point to the normal melting point of nitrogen and beyond. It curves, but its slope is positive, according to the densities given in the problem. The solid/gas equilibrium line extends downward from the triple point. Solids are always more dense than gases, so the slope of this line is positive.

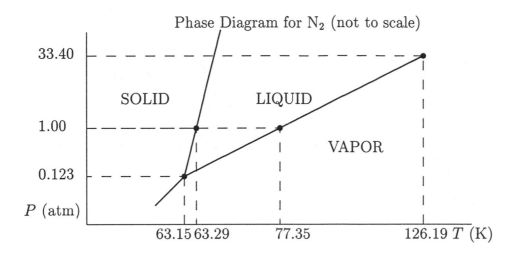

Phase Diagram for N$_2$ (not to scale)

5-48 (a) The phase diagram of H$_2$O (text Figure 5.21c) confirms that at equilibrium at room conditions (the starting conditions in the process described in the problem), water is a liquid.

(b) At 400 K and 1 atm, H$_2$O is a gas, according to the phase diagram.

(c) Although the water starts out as a liquid and ends as a gas in the process described, *no phase transition occurs*. The water is taken into the supercritical region by the changes described. The change from liquid to supercritical fluid is smooth and gradual and the subsequent change from supercritical fluid to gas is also continuous. There is no abrupt change in density or other physical properties and therefore no phase transition.

5-50 (a) The vapor pressure of solid hydrogen, and liquid hydrogen *both* equal the external pressure on the system at the triple point of hydrogen. The answer is 0.069 atm.

(b) The pressure is maintained at a value below the triple-point pressure of hydrogen. Hence the solid hydrogen converts directly to the gas; it sublimes.

5-52 The tube must contain 0.235 g of ammonia for every cm^3 of volume. If it contains more ammonia than this, the ammonia will be liquid and will become supercritical above 132.23°C without anything to be seen from the outside. If it contains less ammonia, the ammonia will be gaseous below the critical temperature and become supercritical above that temperature, again with nothing to be seen. The interior radius of the tube is $5.0 - 4.20 = 0.80$ mm $= 0.080$ cm. The height of the tube is 15.5 cm so that its volume is

$$V = \pi r^2 h = \pi \, (0.080)^2 \text{ cm}^2 \times 15.5 \text{ cm} = 0.3116 \text{ cm}^3$$

The required mass of ammonia is

$$m = dV = (0.235 \text{ g cm}^{-3}) \times (0.3116 \text{ cm}^3) = 0.0732 \text{ g}$$

5-54 The point of the problem is not the calculations, which involve mere substitution into the formula, but the large differences in the diffusive displacement of molecules in the three phases. Let Δr equal the root-mean-square displacement. Then: Then, (a) For $O_2(g)$:

$$\Delta r = \sqrt{6D\Delta t} = \sqrt{6 \, (2.1 \times 10^{-5} \text{ m}^2 \text{ s}^{-1}) \, 3600 \text{ s}} = 0.67 \text{ m}$$

(b) For $H_2O(g)$:

$$\Delta r = \sqrt{6D\Delta t} = \sqrt{6 \, (2.26 \times 10^{-9} \text{ m}^2 \text{ s}^{-1}) \, 3600 \text{ s}} = 0.0070 \text{ m}$$

(c) For $Na(s)$:

$$\Delta r = \sqrt{6D\Delta t} = \sqrt{6 \, (5.8 \times 10^{-13} \text{ m}^2 \text{ s}^{-1}) \, 3600 \text{ s}} = 0.00011 \text{ m}$$

5-56 Compute the ratio a/b for the four substances

$$\frac{a}{b} = \frac{1.390}{0.03913} = 35.52 \text{ L atm mol}^{-1} \quad \text{for } N_2$$

$$\frac{a}{b} = \frac{0.2444}{0.02661} = 9.185 \text{ L atm mol}^{-1} \quad \text{for } H_2$$

$$\frac{a}{b} = \frac{6.714}{0.05636} = 119.1 \text{ L atm mol}^{-1} \quad \text{for } SO_2$$

$$\frac{a}{b} = \frac{3.667}{0.04081} = 89.86 \text{ L atm mol}^{-1} \quad \text{for } HCl$$

The strength of the attractive forces goes as the magnitude of a/b so the ranking is $SO_2 > HCl > N_2 > H_2$.

5-58 Because the density (mass per unit volume) of the water increases as it is heated from 0.0 to 4.0°, its volume per unit mass decreases. The mass of a sample is not changed by heating, so the volume of the water must decrease; the coefficient of thermal expansion is negative in this range of temperature. Positive coefficients of thermal expansion are far more common than negative ones for solids and liquids. The coefficients of thermal expansion of gases are always positive.

5-60 Assuming that the tungsten vapor is an ideal gas (a very good assumption at this exceedingly low pressure) allows computation of the molar volume:

$$\frac{V}{n} = \frac{RT}{P} = \frac{(0.08206 \text{ L atm mol}^{-1}\text{K}^{-1})(1273 \text{ K})}{2 \times 10^{-25} \text{ atm}} = 5.2 \times 10^{26} \text{ L mol}^{-1}$$

This is the volume per mole; the volume per atom is this number divided by Avogadro's number:

$$\left(\frac{5.2 \times 10^{26} \text{ L}}{1\text{mol}}\right) \times \left(\frac{1 \text{ mol}}{6.022 \times 10^{23} \text{ atoms}}\right) = 9 \times 10^2 \frac{\text{L}}{\text{atom}}$$

5-62 The chemical amount of air that was present in the 6.00-L portion of air mixed with the vapors of the unknown can be computed because its physical state after purification is fully described

$$n_{\text{air}} = \frac{PV}{RT} = \frac{(1.000 \text{ atm})(3.75 \text{ L})}{(0.08206 \text{ L atm mol}^{-1}\text{K}^{-1})(223.15 \text{ K})} = 0.2048 \text{ mol}$$

Now, compute the pressure that this chemical amount of air exerted as part of the 6.00-L mixture

$$P_{\text{air}} = \frac{n_{\text{air}}RT}{V}$$

$$P_{\text{air}} = \frac{(0.2048 \text{ mol})(0.08206 \text{ L atm mol}^{-1}\text{K}^{-1})(298.15 \text{ K})}{6.00 \text{ L}} = 0.835 \text{ atm}$$

But the total pressure above the unknown was 0.980 atm. By Dalton's law

$$P_{\text{unknown}} = 0.980 - 0.835 = 0.145 \text{ atm}$$

5-64 If the pressure inside the lighter is not to exceed 1 atm, then the butane must be a gas. Pressurization is the only way to keep the butane a liquid at room temperature, because room temperature exceeds its normal boiling point. Estimating the amount of gaseous butane in a lighter with a storage volume of 10 mL requires substitution in the ideal gas equation and solving for n. The conditions are: $P = 1$ atm, $V = 0.01$ L, and $T = 298$ K. Doing the arithmetic gives $n = 4.1 \times 10^{-4}$ mol of butane which amounts to 0.024 g of butane (the $\mathcal{M}$ of butane is 58.1 g mol^{-1}). This is about 1/200 of the butane in a standard lighter.

5-66

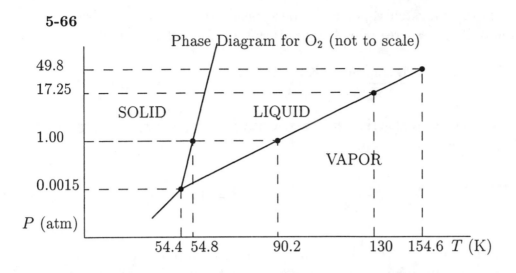

Phase Diagram for O_2 (not to scale)

5-68 (a) A 400 atm pressure would lower the melting point of ice to about $-4°C$.

(b) The pressure exerted by the blade of the skate is

$$P = \frac{F}{A} = \frac{ma}{A} = \frac{(75 \text{ kg}) (9.8 \text{ m s}^{-2})}{8.0 \times 10^{-5} \text{ m}^2} = 9.2 \times 10^6 \text{ Pa}$$

This pressure equals about 90 atm. Therefore, ice at $-5°C$ does not melt under the pressure of the blade.

5-70 HBr has stronger attractive forces than HCl because of its larger molar mass. HF also has stronger attractive forces but for a different reason: it forms hydrogen bonds. Stronger forces drive up the characteristic temperature to form a gas from a liquid.

5-72

$$\frac{n}{V} = \frac{P}{RT} = \frac{7.0 \times 10^{-9} \text{ atm}}{(0.08206 \text{ L atm mol}^{-1}\text{K}^{-1})(2773 \text{ K})}$$

$$= 3.08 \times 10^{-11} \text{ mol L}^{-1}$$

$$= 3.08 \times 10^{-14} \text{ mol cm}^{-3}$$

Multiplying by Avogadro's number gives

$$(3.08 \times 10^{-14} \text{ mol cm}^{-3})(6.022 \times 10^{23} \text{ atoms mol}^{-1}) = 1.9 \times 10^{10} \text{ atoms cm}^{-3}$$

5-74 The total chemical amount of the original mixture is found from the ideal gas law:

$$n_{\text{tot}} = \frac{(0.980 \text{ atm})(6.00 \text{ L})}{(0.08206 \text{ L atm mol}^{-1}\text{K}^{-1})(298.15 \text{ K})} = 0.2403 \text{ mol}$$

The chemical amount of the air is found in the same way

$$n_{\text{air}} = \frac{(1.000 \text{ atm})(3.75 \text{ L})}{(0.08206 \text{ L atm mol}^{-1}\text{K}^{-1})(223.15 \text{ K})} = 0.02048 \text{ mol}$$

The difference, $0.2403 - 0.2048 \text{ mol} = 0.0355 \text{ mol}$, is the chemical amount of unknown compound in the original vapor. Its partial pressure is

$$P_{\text{i}} = X_{\text{i}}P_{\text{tot}} = \frac{0.0355 \text{ mol}}{0.2403 \text{ mol}} \times 0.980 \text{ atm} = 0.145 \text{ atm}$$

Chapter 6

Solutions

6-2 We convert the given amount of C_2H_5OH from grams to moles and the volume of blood from deciliters to liters:

$$\frac{0.1\,\text{g }C_2H_5OH}{1\,\text{dL}} \times \left(\frac{10\,\text{dL}}{1\,\text{L}}\right) \times \left(\frac{1\,\text{mol }C_2H_5OH}{46.1\,\text{g }C_2H_5OH}\right) = 0.02\,\text{mol L}^{-1}$$

6-4 (a) Let us suppose that we have exactly 1.0000 L of the solution. We then must have 205.0 g of acetic acid and 820.0 g of water. The total mass of the solution is 1025.0 g. Density $= \dfrac{1025.0\,\text{g}}{1000\,\text{cm}^3} = 1.0250\,\text{g cm}^{-3}$

(b) We use the molar masses of acetic acid and water to convert the two masses to chemical amounts. They are $205.0\,\text{g}/60.05\,\text{g mol}^{-1} = 3.4136\,\text{mol}$ acetic acid and $820.0\,\text{g}/18.015\,\text{g mol}^{-1} = 45.517\,\text{mol}$ water. The molarity of the acetic acid is the number of moles of acetic acid per liter of solution. Our 1.000 L of solution contains 3.4136 mol of acetic acid, so the answer is 3.414 M. The molality of the acetic acid is the number of moles of acetic acid per kilogram of solvent. We have 3.4136 mol of acetic acid dissolved in 0.8200 kg of water. The molality of the acetic acid is therefore

$$3.4136\,\text{mol}/0.8200\,\text{kg} = 4.163\,\text{mol kg}^{-1}$$

The mole fraction of acetic acid is

$$\frac{3.4136\,\text{mol}}{3.4136\,\text{mol} + 45.517\,\text{mol}} = 0.06976$$

and the mass percentage of the acetic acid is

$$\frac{205.0\,\text{g}}{1025.0\,\text{g}} \times 100\% = 20.00\%$$

(c) The molarity of the solute water in this solution is the number of moles of water per liter of solution. Our 1.0000 L of solution contains 45.517 mol of water so the answer is 45.52 M.

The molality of the water is 45.517 mol/0.2050 kg acetic acid = 222.03 mol kg^{-1}. The mole fraction of water is

$$\frac{45.517 \text{ mol}}{3.4136 + 45.517 \text{ mol}} = 0.9302$$

Finally, the mass percent of the water is (820.0 g/1025.0 g) × 100% = 80.00%

6-6 One liter has a mass of 1171 g and it contains 1.241 mol AgNO$_3$. The mass of this amount of AgNO$_3$ is

$$(1.241 \text{ mol AgNO}_3)\left(169.873 \text{ g mol}^{-1}\right) = 210.8 \text{ g AgNO}_3$$

so the balance (960 g) must be water. The molality is

$$\frac{1.241 \text{ mol AgNO}_3}{0.960 \text{ kg H}_2\text{O}} = 1.29 \text{ molal solution}$$

6-8 The mass of the 1.00 L of water-in-methane solution is 780 g. The molecular formula of methane is CH$_4$ so $\mathcal{M} = 16.04$ g mol^{-1}; water has $\mathcal{M} = 18.02$ g mol^{-1} where $\mathcal{M}$ stands for molar mass. Let x equal the mass in grams of water in the solution and y equal the mass in grams of methane. Then clearly

$$x + y = 780 \text{ g}$$

Also, according to the definition of mole fraction

$$6.0 \times 10^{-5} = \frac{x/18.02}{(x/18.02) + (y/16.04)}$$

Solving these two simultaneous equations for x gives $x = 0.053$ g. This mixture is quite dilute in the solute water. Some algebraic labor can therefore be saved by omitting the term in x in the denominator; it is quite small compared to the term in y.

6-10 (a) Imagine a sample of the perchloric acid solution that weighs exactly 1000 g. This solution contains 600 g of HClO$_4$ ($\mathcal{M} = 100.46$ g mol^{-1}) which is equivalent to 5.973 mol of HClO$_4$. Thus, this solution contains 5.973 mol of solute per 1000 g of solution. But it also contains 9.20 mol of solute per 1000 mL of solution, because its molarity is given as 9.20 M. Using proportions, we conclude the solution must contain 5.973 mol per $(5.973/9.20) \times 1000$ mL $=$ 649.2 mL. Because 1000 g of the solution is equal to 649.2 mL, the density of the solution is $1000/649.2 = 1.54$ g mL^{-1}.

(b) To prepare 1.00 L of a solution that contains 1.00 mol of the solute, we need 1.00 mol of the $HClO_4$. The 9.20 M solution is quite concentrated: 1000 mL of it supplies 9.20 mol of $HClO_4$. It follows that $1000/9.20 = 108.7$ mL of it supplies 1.00 mol of solute.

6-12 The 0.400 L sample contains 0.400 L $\times$ 0.0700 mol L^{-1} $= 0.0280$ mol HNO_3; the 0.800 L sample contains 0.800 L $\times$ 0.0300 mol L^{-1} $= 0.0240$ mol HNO_3 so that the solution that results, after mixing, contains the sum of 0.0240 and 0.0280 or 0.0520 mol HNO_3. The volume of this solution is 1.200 L; its concentration is 0.0520 mol/1.200 L or 0.0433 M. Note the assumption that the volumes of the two solutions are additive. This is not always defensible, but is justified in the case of dilute aqueous solutions such as these two.

6-14 Net ionic equations omit all ions not specifically reacting:
(a) $Ba^{2+}(aq) + SO_4^{2-}(aq) \rightarrow BaSO_4(s)$
(b) $3\,Cl_2(g) + 6\,OH^-(aq) \rightarrow ClO_3^-(aq) + 5\,Cl^-(aq) + 3\,H_2O(l)$
(c) $Hg_2^{2+}(aq) + 2\,I^-(aq) \rightarrow Hg_2I_2(s)$
(d) $3\,OCl^-(aq) + I^-(aq) \rightarrow IO_3^-(aq) + 3\,Cl^-(aq)$

6-16 $\left(\dfrac{2200 \times 10^3 \text{ g}}{504.3 \text{ g mol}^{-1}}\right) = 4.36 \times 10^3$ mol $Ca_5(PO_4)_3F$

4.63×10^3 mol $Ca_5(PO_4)_3F \left(\dfrac{3 \text{ mol } H_3PO_4}{1 \text{ mol } Ca_5(PO_4)_3F}\right) = 1.31 \times 10^4$ mol H_3PO_4

$\dfrac{1.31 \times 10^4 \text{ mol}}{6.3 \text{ mol L}^{-1}} = 2.1 \times 10^3$ L

6-18 At 20°C, and a pressure of 0.970 atm, the chemical amount of NO is readily calculated by assuming ideal gas behavior and writing $n = PV/RT$. It is 0.2016 mol. This amount of $NO(g)$ requires $\frac{6}{4} \times 0.2016 = 0.3024$ mol of $NaNO_2$ for its generation, acording to the balanced equation.

$$\frac{0.3024 \text{ mol}}{0.646 \text{ mol L}^{-1}} = 0.468 \text{ L solution}$$

6-20 (a) $2\,NaOH + H_2SO_3 \rightarrow 2\,H_2O + Na_2SO_3$.
This reaction involves the base sodium hydroxide, the acid sulfurous acid, and the salt sodium sulfite.

(b) $Ca(OH)_2 + 2\,C_6H_5COOH \rightarrow 2\,H_2O + Ca(C_6H_5COO)_2$
This reaction involves the base calcium hydroxide, the acid benzoic acid, and the salt calcium benzoate.

(c) $Pb(OH)_2 + H_2SO_4 \rightarrow 2 H_2O + PbSO_4$.

This reaction involves the base lead hydroxide, sulfuric acid, and the salt lead(II) sulfate.

(d) $Cu(OH)_2 + 2 HCl \rightarrow 2 H_2O + CuCl_2$.

This reaction involves the base copper(II) hydroxide, hydrochloric acid, and the salt copper(II) chloride.

6-22 The salt comes from sodium hydroxide and sulfuric acid, so it is sodium sulfate (Na_2SO_4). The answer sodium hydrogen sulfate ($NaHSO_4$) is marginally possible, but the context suggests the complete neutralization of the sulfuric acid.

6-24 (a)

$$PCl_5 + 4 H_2O \rightarrow H_3PO_4 + 5 HCl$$

(b) By the ideal gas law, the chemical amount contained in 1.22 L of any gas confined at this temperature (488.15 K) and pressure is 0.0293 mol. The 0.0293 mol of PCl_5 generates $5 \times 0.0293 = 0.146$ mol of HCl and 0.0293 mol of H_3PO_4. Both acids are collected in enough water so that the final volume of the mixed solution is 697 mL. The concentration of the H_3PO_4 is 0.0293 mol/0.697 L = 0.0420 M, and the concentration of the HCl is five times this, or 0.210 M.

6-26 Hydrochloric acid and ammonia react in a 1:1 molar ratio and we assume that bases other than ammonia are absent. Therefore, the chemical amount of hydrochloric acid required to bring the ammonia-containing cleaning solution to the end-point equals the chemical amount of ammonia originally present. The chemical amount of HCl is

$$0.8381 \text{ mol L}^{-1} \times 0.02318 \text{ L} = 0.01943 \text{ mol}$$

This much ammonia was present in 50.0 mL of the cleaning solution. Therefore, the concentration of ammonia in the cleaning solution is

$$\frac{0.01943 \text{ mol}}{0.0500 \text{L}} = 0.389 \text{ M}$$

6-28 (a) Nitrogen changes oxidation state from +4 in N_2O_4 to +3 in NOCl and +5 in KNO_3.

(b) Sulfur changes oxidation state from −2 in H_2S to +6 in SF_6. Oxygen changes oxidation state from +1 in O_2F_2 to 0 in O_2.

(c) Phosphorus changes oxidation state from +5 in $POBr_3$ to +2 in PO. Magnesium changes from 0 in Mg to +2 in $MgBr_2$.

(d) Some of the chlorine changes oxidation state from -1 in BCl_3 to 0 in Cl_2. Sulfur changes from $+4$ in SF_4 to $+2$ in SCl_2.

6-30 In the equation the iodine is in the $+5$ oxidation state on the left side of the equation and ends up in the 0 oxidation state on the right side. Carbon is in the $+2$ oxidation state on the left side and goes to the $+4$ oxidation state on the right. All oxygen stays in the -2 oxidation state throughout the reaction. I_2O_5 is reduced. CO is oxidized.

6-32 The sample of As_2O_3 ($\mathcal{M} = 197.84$ g mol^{-1}) contains 1.097×10^{-3} mol of As_2O_3. Upon dissolution it gives $2 \times (1.097 \times 10^{-3})$ or 2.194×10^{-3} mol of H_3AsO_3 because there are two As atoms per As_2O_3 formula unit. The titration consumes twice this chemical amount of Ce^{4+} because these two species react in a 2:1 molar ratio, according to the balanced equation. This amount of Ce^{4+} ion (4.39×10^{-3} mol) is delivered by the addition of 0.02147 L of solution during the titration, so the molarity of the Ce^{4+} ion in the titrating solution is 0.204 M.

6-34 The chemical amount of maleic acid is 0.06203 mol, and that of diethyl ether is 1.349 mol. The mole fraction of maleic acid (the solute) is then

$$X_{\text{solute}} = \frac{0.06203 \text{ mol}}{1.349 + 0.06203 \text{ mol}} = 0.04396$$

The vapor pressure lowering is

$$\Delta P = -P° X_{\text{solute}} = -(0.8517 \text{ atm})(0.04396) = -0.0374 \text{ atm}$$

so the vapor pressure of ether above the solution is

$$P = 0.8517 - 0.0374 \text{ atm} = 0.8143 \text{ atm}$$

6-36 The chemical amount of anthracene is

$$\frac{2.62 \text{ g}}{178.23 \text{ g mol}^{-1}} = 0.0147 \text{ mol}$$

Its molality is

$$\frac{0.0147 \text{ mol}}{0.1000 \text{ kg}} = 0.147 \text{ mol kg}^{-1}$$

The constant K_b is then

$$K_b = \frac{\Delta T}{m} = \frac{0.41 \text{ K}}{0.147 \text{ mol kg}^{-1}} = 2.8 \text{ K kg mol}^{-1}$$

6-38 The dissolution of the compound of indium and chlorine in tin(IV) chloride raises the boiling point of the solvent. We assume that this solute is nonvolatile. Then, the increase in the boiling point gives the molality of the solution:

$$m = \frac{\Delta T}{K_b} = \frac{2.2 \text{ K}}{9.43 \text{ K kg mol}^{-1}} = 0.233 \text{ mol kg}^{-1}$$

2.60 g dissolved in 50.0 g of solvent has the same molality as 52.0 g dissolved in 1.00 kg of solvent. Thus 52.0 g is equivalent to 0.233 mol, and

$$\mathcal{M} = \frac{52.0 \text{ g}}{0.233 \text{ mol}} = 223 \text{ g mol}^{-1}$$

The probable formula is $InCl_3$.

6-40 The change in the freezing point of the solution ΔT is $937 - 962 = -25°C = -25$ K. Inserting this and the K_f value into the freezing point depression equation gives a molality of 0.231 mol kg^{-1}. But this solution contains

$$12 \text{ g solute}/0.562 \text{ kg solvent} = 21.35 \text{ g kg}^{-1}$$

of solute, according to the statement of the problem. The molar mass of the unknown is therefore $21.35/0.231 = 92$ g mol^{-1}.

6-42 There is 1/8 kg = 125 g NaCl per kg of solvent. This corresponds to 2.14 mol NaCl, so $m = 2 \times 2.14 = 4.28$ mol kg^{-1} because each NaCl gives two ions in solution.

$$\Delta T = -K_f\, m = -(1.86 \text{ K kg mol}^{-1})(4.28 \text{ mol kg}^{-1}) = -8.0 \text{ K}$$

This gives a freezing point of $-8.0°C$, or $18°F$.

6-44 The solution of ethanol in sulfuric acid has a molality of 0.050 mol kg^{-1} because 2.3 g of ethanol ($\mathcal{M} = 46$ g mol^{-1}) equals 0.050 mol of ethanol. A solution of this molality would be expected to depress the freezing point of the sulfuric acid by 6.12 K kg mol^{-1} × 0.050 mol kg^{-1} = 0.306 K. The actual depression is 0.92 K, which is nearly exactly three times as great. Therefore, the ethanol apparently reacts to form three particles per molecule upon dissolution in sulfuric acid. This reaction has been studied. It is:

$$C_2H_5OH + 2H_2SO_4 \rightarrow C_2H_5HSO_4 + HSO_4^- + H_3O^+$$

6-46 $c = \dfrac{\pi}{RT} = \dfrac{0.0319 \text{ atm}}{(0.08206 \text{ L atm mol}^{-1} \text{ K}^{-1})(293.15 \text{ K})} = 0.001326 \text{ mol L}^{-1}$

$$\mathcal{M} = \frac{23.7 \text{ g L}^{-1}}{0.001326 \text{ mol L}^{-1}} = 1.79 \times 10^4 \text{ g mol}^{-1}$$

6-48

$$\pi = \rho gh = (0.789 \times 10^3 \text{ kg m}^{-3})(9.80665 \text{ m s}^{-2})(0.263 \text{ m})$$

$$= 2.035 \times 10^3 \text{ Pa} = 2.01 \times 10^{-2} \text{ atm}$$

$$c = \pi/RT = 8.35 \times 10^{-4} \text{ mol L}^{-1}$$

One liter contains 12.5 g of protein, or 8.35×10^{-4} mol. Dividing gives the molar mass:

$$\mathcal{M} = 12.5 \text{ g}/8.35 \times 10^{-4} \text{ mol} = 1.50 \times 10^4 \text{ g mol}^{-1}$$

6-50 (a) The partial pressures in air are $P_{N_2} = 0.78$ atm and $P_{O_2} = 0.21$ atm. Thus, from Henry's law

$$P_i = k_i X_i$$

we have

$$X_{N_2} = \frac{P_{N_2}}{k_{N_2}} = 0.78 \text{ atm}/8.57 \times 10^4 \text{ atm} = 9.10 \times 10^{-6}$$

$$X_{O_2} = 0.21 \text{ atm}/4.34 \times 10^4 \text{ atm} = 4.84 \times 10^{-6}$$

One liter of water contains 55.5 mol water. Because the mole fractions of gases dissolved are so small, we can write

$$n_{N_2} = X_{n_2}(n_{N_2} + n_{O_2} + n_{H_2O}) \approx X_{N_2} n_{H_2O}$$

$$= (9.10 \times 10^{-6})(55.5 \text{ mol}) = 5.1 \times 10^{-4}$$

$$[N_2] = 5.1 \times 10^{-4} M$$

In the same way, we calculate

$$[O_2] = 2.7 \times 10^{-4} M$$

(b) For a given partial pressure, less helium dissolves in the diver's blood than nitrogen because helium has an intrinsically lower solubility (larger Henry's law constant) in water than nitrogen. The substitution of helium for nitrogen in breathing mixtures for divers means that a smaller amount of gas dissolves in the divers' blood and moderates the risk of "bends."

6-52 The volume of methane that is expelled by boiling the methane-in-benzene solution is related to its chemical amount by the ideal gas law. At 0°C and 1.00 atm, 0.510 L of gas amounts to 0.0228 mol of gas. This amount of methane was present in 1.00 kg of benzene ($\mathcal{M} = 78.114$ g mol^{-1}), which is 12.80 mol of benzene. The mole fraction of dissolved methane therefore must have been

$$\frac{0.0228}{12.80 + 0.0228} \approx \frac{0.0228}{12.80} = 1.78 \times 10^{-3}$$

This fraction of dissolved methane was forced into the benzene by a pressure of 1.00 atm of methane. Henry's law for this case is

$$P_{methane} = kX_{methane}$$

We know the pressure and mole fraction of the methane, so k is easily calculated. It is 563 atm.

6-54 The vapor pressure is simply the vapor pressure that the component would have, if it were pure, multiplied by its mole fraction in the solution. Therefore, the vapor pressure of the toluene above this solution is

$$\left(\frac{0.900}{0.400 + 0.900}\right)(0.534 \text{ atm}) = 0.370 \text{ atm}$$

at this temperature, and the vapor pressure of the benzene is 0.412 atm, by a similar calculation. The total pressure of the vapors above the solution is 0.782 atm, the sum of the partial pressures of the two volatile components of the solution. The mole fraction of benzene is

$$0.412 \text{ atm}/0.782 \text{ atm} = 0.527$$

6-56 (a) The chemical amount of benzene ($\mathcal{M} = 78.11$ g mol^{-1}) in 50.0 g is 0.640 mol. The chemical amount of n-hexane ($\mathcal{M} = 86.18$ g mol^{-1}) is 0.580 mol. The mole fraction of benzene in the solution is therefore $\frac{0.640}{0.640 + 0.580} = 0.525$.

(b) The total vapor pressure of the solution is the vapor pressure of the benzene added to the vapor pressure of the n-hexane. These two vapor pressures are computed using Raoult's law. They are $P_{benzene} = 0.0711$ atm and $P_{n-hexane} = 0.101$ atm. The total vapor pressure is therefore 0.172 atm.

(c) The mole fraction of benzene in the vapor is $0.0711/0.172 = 0.413$.

6-58 (a) $Ba(NO_3)_2$ (in NH_3) + 2 AgBr (in NH_3) → $BaBr_2(s)$ +2 $AgNO_3$ (in NH_3)

(b) $(0.215 \text{ L}) (0.076 \text{ mol L}^{-1}) = 0.01634 \text{ mol } Ba(NO_3)_2$

$0.01634 \text{ mol } Ba(NO_3)_2 \left(\frac{2 \text{ mol AgBr}}{1 \text{ mol } Ba(NO_3)_2}\right) = 0.03268 \text{ mol AgBr}$

$\frac{0.03268 \text{ mol}}{0.50 \text{ mol L}^{-1}} = 0.065 \text{ L} = 65 \text{ mL}$

(c) $(0.01634 \text{ mol } BaBr_2) \times 297.14 \text{ g mol}^{-1} = 4.86 \text{ g}$

6-60 It requires $0.0262 \text{ L} \times 0.1359 \text{ mol L}^{-1} = 3.56 \times 10^{-3}$ mol of thiosulfate ion to titrate the I_3^- ion. From the stoichiometry of the two reactions that are given in the problem there must have been exactly one-half this chemical amount of I_3^- or 1.78×10^{-3} mol of I_3^- present and the identical chemical amount of $O_3(g)$ had to be in the original mixture to generate the I_3^- ion. The ideal gas equation gives the total number of moles of gases in the original mixture. It is

$$n = \frac{PV}{RT} = \frac{(0.993 \text{ atm })(53.2 \text{ L})}{(0.08206 \text{ L atm mol}^{-1} \text{K}^{-1})(291.15 \text{ K})} = 2.211 \text{ mol}$$

The mole fraction of the ozone is the number of moles of ozone detected in the analysis divided by the total number of moles of gases in the mixture. It is 8.05×10^{-4} .

6-62 If the mole fraction of the NaI is 0.390, then there is 39.0 mol of NaI in the solution for every 61.0 mol of water. Converting this amount of water to kilograms (using the molar mass of water) reveals that the solution has 39.0 mol of NaI per 1.0989 kg of water or 35.5 mol of NaI per 1.00 kg of water. If we assume NaI is fully dissociated into ions, the total molality is twice this, or 71.0 mol kg^{-1}. Inserting this molality into $\Delta T = K_b m$ with $K_b = 0.512$ K kg mol^{-1} gives $\Delta T = 36$ K. The experimental ΔT is 44 K, which is somewhat higher.

6-64 The apparent molality of the solution is computed from the freezing point depression. It is

$$m = -\frac{\Delta T}{K_f} = -\frac{-0.99 \text{ K}}{34.3 \text{ K kg mol}^{-1}} = 0.0289 \text{ mol kg}^{-1}$$

This means that the 1.36 g of solid mercury(I) chloride that is dissolved in 100 g of the solvent must be equal to 0.00289 mol of mercury(I) chloride. From this we find that the molar mass of mercury(I) chloride is 471 g mol^{-1}. If mercury(I) chloride ionized in this solvent, then its apparent molar mass would be *less* than 236 g mol^{-1}, which is the mass of the HgCl formula unit. Instead, the apparent molar mass is almost exactly twice this number. Instead of ionizing to give more particles the "HgCl" in this solution clusters together in dimers; it exists as Hg_2Cl_2.

6-66 The molality of the required anti-freeze solution of ethylene glycol is

$$m = -\frac{\Delta T}{K_f} = -\frac{(-5.0)}{1.86} = 2.69 \text{ mol kg}^{-1}.$$

Such a solution has 2.69 mol of CH_2OHCH_2OH ($\mathcal{M} = 62.07$ g mol^{-1}) dissolved in 1000 g of water. This chemical amount of ethylene glycol equals 166.9 g of ethylene glycol. There is thus 166.9 g of ethylene glycol for every 1166.9 g of solution. The percentage by mass of the ethylene glycol in the solution is $166.9/1166.9 \times 100\% = 14.3\%$. Note the assumption that ethylene glycol neither dissociates into two or more particles nor associates in aqueous solution.

6-68 The vapor pressure above the first beaker is higher than that above the second beaker. Thus as time goes on there will be net evaporation from the first and condensation into the second until the two concentrations become equal, at which point the vapor pressures will also be equal.

First beaker contains $(0.400 \text{ L})(0.100 \text{ mol L}^{-1}) = 0.0400$ mol

Second beaker contains $(0.200 \text{ L})(0.250 \text{ mol L}^{-1}) = 0.0500$ mol

Total $= 0.0900$ mol; total volume $= 600$ mL

Final concentration $= \dfrac{0.0900 \text{ mol}}{0.600 \text{ mol L}} = 0.150$ M

Volume in 1 $= \dfrac{0.0400 \text{ mol}}{0.150 \text{ mol L}^{-1}} = 267$ mL

Volume in 2 $= \dfrac{0.0500 \text{ mol}}{0.150 \text{ mol L}^{-1}} = 333$ mL

6-70 First we compute the osmotic pressure of the solution:

$$
\begin{aligned}
\pi &= \tfrac{n}{V}RT = (0.010 \text{ mol L}^{-1})(0.08206 \text{ L atm mol}^{-1}\text{K}^{-1})(696.15 \text{ K}) \\
&= 0.571 \text{ atm} = 5.79 \times 10^4 \text{ Pa} \\
\pi &= \rho g h \\
h &= \pi/\rho g = \frac{5.79 \times 10^4 \text{ Pa}}{(11.4 \times 10^3 \text{ kg m}^{-3})(9.80665 \text{ m s}^{-2})} = 0.52 \text{ m}
\end{aligned}
$$

6-72 According to the data in problem 6-54, the vapor pressure of toluene is 0.534 atm at 90°C, and the vapor pressure of benzene is 1.34 atm. In order for the solution to boil, the total pressure above it must equal 1.00 atm. This total pressure is the sum of the pressures of the two components, each of which is given by Raoult's law. If we let the mole fraction of the toluene in solution equal x, then the mole fraction of the benzene is $1 - x$, and we can write

$$0.534x + 1.34(1 - x) = 1.00$$

Solving this equation for x gives 0.422 as the mole fraction of the toluene.

6-74 The mass of 1.000 mol of Na_2SO_4 is 142.04 g. The additional mass, $322.2 - 142.04 = 180.2$ g, must be water of hydration. Its chemical amount is

$$\frac{180.2 \text{ g}}{18.02 \text{ g mol}^{-1}} = 10.0 \text{ mol H}_2\text{O}$$

so the chemical formula of the solid hydrate is $Na_2SO_4 \cdot 10H_2O$.

Chapter 7

Thermodynamic Processes and Thermochemistry

7-2 The work done in a change of volume at constant pressure is $w = -P_{ext}\Delta V$. The change in the volume is $V_2 - V_1 = 800 - 150 = 650$ mL $= 0.650$ L, and P is 0.98 atm so $w = -0.64$ L atm. 1 L atm is 101.325 J. The work is therefore -65 J.

7-4 We can write

$$-mg\Delta h = c_s m \Delta T$$

The m cancels out. Before inserting the given values into this equation, we convert c_s to units of J K^{-1}kg^{-1} to assure that the units will cancel out correctly. We then have:

$$-(9.81 \text{ m s}^{-2})(-100 \text{ m}) = (4180 \text{ J K}^{-1}\text{kg}^{-1})\Delta T$$

From this expression $\Delta T = +0.235$ K.

7-6 The molar heat capacities are the specific heats multiplied by the molar masses

$$c_P = \mathcal{M}c_s.$$

Analysis of the units confirms this relationship. The molar heat capacities (at constant pressure) for the gaseous halogens come out to be:

	c_s(J K^{-1}g^{-1})	$\mathcal{M}$(g mol^{-1})	c_P(J K^{-1}mol^{-1})
F$_2$	0.824	38.0	31.3
Cl$_2$	0.478	70.9	33.9
Br$_2$	0.225	159.8	36.0
I$_2$	0.145	253.8	36.8

An increasing trend is established among these elements. The molar heat capacity of solid astatine probably lies between 37 and 38 J K^{-1}mol^{-1}.

7-8 The rule of Dulong and Petit states that the molar heat capacities (c_P) of the (solid) metallic elements equal approximately 25 J K^{-1}mol^{-1}. Because $c_P = \mathcal{M}c_s$ we merely divide a metallic element's c_P by its molar mass to estimate its specific heat. The answers are: 0.491, 0.359, and 0.232 J K^{-1}g^{-1} for V, Ga and Ag, respectively.

7-10 (a) We identify the battery (not the toy truck) as the system of interest and apply the first law to it. The battery performs work on the surroundings so $w = -117.0$ J. It also absorbs negative heat (evolves heat to the surroundings) so $q = -3.0$ J. By the first law, $\Delta E = -120.0$ J.

(b) Recharging the battery puts it back the way it was before it was discharged. The final state is indistinguishable from the original, state, so $\Delta E_{\text{overall}} = 0$. But

$$\Delta E_{\text{overall}} = 0 = \Delta E_{\text{discharge}} + \Delta E_{\text{charge}} = (q + w)_{\text{discharge}} + (q + w)_{\text{charge}}$$

Substituting the known quantities in joules gives

$$0 = -3.0 + (-117.0) + q_{\text{charge}} + 210.0 \quad \text{which means} \quad q_{\text{charge}} = -90.0 \text{ J}$$

The sign of q during the charge cycle is negative because the battery evolves heat to the surroundings as it is charged.

7-12 We assume that the water and zinc are thermally isolated from the rest of the world. Then,

$$q_{\text{water}} + q_{\text{zinc}} = 0$$

$$(c_{s,\text{water}} m_{\text{water}} \Delta t_{\text{water}}) + (c_{s,\text{zinc}} m_{\text{zinc}} \Delta t_{\text{zinc}}) - 0$$

The Δt of the zinc is $(t_f - 20.0)$ and the Δt of the water is $(t_f - 100.0)$. Substituting these quantities, the specific heats, and the masses gives:

$$(4.22(200.0)(t_f - 100.0)) + (0.389(60.0)(t_f - 20.0)) = 0$$

Solving gives $t_f = 97.85°$C.

7-14 This problem can be solved by setting up an equation like the one used to solve problem 7-12. One can also use the rule derived in problem 7-13. In the special case of mixing equal masses of substances 1 and 2 at different temperatures:

$$\frac{c_{s1}}{c_{s2}} = -\frac{\Delta t_2}{\Delta t_1}$$

In this problem, this expression becomes

$$0.10 = -\frac{(t_f - 20.0)}{(t_f - 92.0)}$$

The minus signs appear in these two equations because of the convention that a change in a quantity is the final value minus the initial. Solving gives $t_f = 26.5°C$.

7-16 Let m represent the original masses of ice and boiling water, and define the system as the mixture of the ice and boiling water. The q for the system in the insulated container is 0. Define subsystem i as the part of the system that was originally ice at 0°C and subsystem hw as the part of the system that was originally boiling water at 100.0°C. Then

$$q_{ice} + q_{hw} = 0$$

The heat absorbed by the ice as it melts is $333.4m$ J. The ice forms water at 0.00°C as it melts. The melted ice then absorbs heat as it rises in temperature to the "Galen temperature" t_G. This heat equals $4.184 \times m \times (t_G - 0.00)$ J. The heat absorbed by the boiling water as it attains the Galen temperature is $4.184 \times m \times (t_G - 100.0)$ J. Hence

$$333.4m + 4.184m(t_G) + 4.184m(t_G - 100.0) = 0$$

The m divides out of this equation which is then readily solved for t_G. The answer is 10.2°C.

7-18 (a)

$$\text{Initial temperature} = \frac{(2.00 \text{ atm})(100 \text{ L})}{(6.00 \text{ mol})(0.08206 \text{ L atm mol}^{-1}\text{K}^{-1})} = 406 \text{ K}$$

$$\text{Final temperature} = \left(\frac{50 \text{ L}}{100 \text{ L}}\right)(406 \text{ K}) = 203 \text{ K}$$

(b)

$$w = -P\Delta V = -(2.00 \text{ atm})(50 \text{ L} - 100 \text{ L}) = +100 \text{ L atm} = +1.01 \times 10^4 \text{ J}$$

(c)

$$\Delta E = nc_V \Delta T = n(c_P - R)\Delta T$$
$$= (6.00 \text{ mol})(29.3 - 8.315 \text{ J mol}^{-1} \text{ K}^{-1})(203 - 406 \text{ K})$$
$$= -2.56 \times 10^4 \text{ J}$$

(d)

$$q = \Delta E - w = -3.57 \times 10^4 \text{ J}$$

7-20 For an ideal gas, $\Delta E = nc_V \Delta T = 0$ if $\Delta T = 0$ (isothermal). Because $P_{\text{ext}} = 0$, $w = -P_{\text{ext}} \Delta V = 0$

$$q = \Delta E - w = 0 - 0 = 0$$

7-22 (a)

$$w = -P_{\text{ext}} \Delta V = -2.00 \text{ atm}(10.00 - 6.00 \text{ L}) = -8.00 \text{ L atm} = -811 \text{ J}$$

$$\Delta E = q + w = +500 - 811 = -311 \text{ J}$$

(b) $\Delta E = -311$ J because E is a function of state

$$w = \Delta E - q = -311 \text{ J} - 0 = -311 \text{ J}$$

7-24 (a)

$$\frac{-683 \text{ kJ}}{1 \text{ mol Br}_2} \times \frac{1 \text{ mol Br}_2}{2(79.904) \text{ g Br}_2} = -4.27 \text{ kJ g}^{-1}$$

(b)

$$\frac{+472 \text{ kJ}}{4 \text{ mol Fe}_3\text{O}_4} \times \frac{1 \text{ mol Fe}_3\text{O}_4}{231.54 \text{ g Fe}_3\text{O}_4} = +0.510 \text{ kJ g}^{-1}$$

(c)

$$\frac{+806 \text{ kJ}}{2 \text{ mol NaHSO}_4} \times \frac{1 \text{ mol NaHSO}_4}{120.06 \text{ g NaHSO}_4} = +3.36 \text{ kJ g}^{-1}$$

7-26 The reaction "absorbs negative heat" at constant pressure, so its ΔH is -670 J. Then:

$$\left(\frac{-670 \text{ J}}{1.00 \text{ g CuCl}_2} \right) \times \left(\frac{134.452 \text{ g CuCl}_2}{1 \text{ mol CuCl}_2} \right) = -90.0 \times 10^3 \text{ J mol}^{-1}$$

7-28 In this process a liquid freezes, so ΔH of the liquid is negative:

$$\left(\frac{-28.8 \text{ kJ}}{1 \text{ mol NaCl}} \right) \times \left(\frac{1 \text{ mol NaCl}}{58.44 \text{ g NaCl}} \right) \times (56.2 \times 10^3 \text{ g NaCl}) = -2.77 \times 10^4 \text{ kJ}$$

7-30 The ice at 0.0°C melts to form water at 0.0°C. The water from the melted ice has $\Delta T = 0$ (absorbs no heat) and can be omitted in the heat balance. We have

$$q_{\text{ice}} + q_{\text{water}} = 0$$

$$\Delta H_{\text{fus}} m_{\text{ice}} + (c_s m \Delta T)_{\text{water}} = 0$$

$$333 \text{ J g}^{-1} m + 4.18 \text{ J K}^{-1}\text{g}^{-1}(150 \text{ g})(-25 \text{ K}) = 0$$

Solving gives $m_{\text{ice}} = 47$ g.

7-32 The reaction of interest can be constructed as three times the first reaction added to the second reaction. The enthalpy changes combine in the same way as the equations. Thus:

$$\Delta H = \Delta H_2 + 3\Delta H_1 = 461.05 + 3(520.9) = +2023.8 \text{ kJ}$$

7-34 The conversion of monoclinic sulfur to rhombic sulfur can be imagined to proceed by the combustion of monoclinic sulfur to SO_2 followed by the "uncombustion" of SO_2 to rhombic sulfur. Therefore $\Delta H = -9.376 - (-9.293) = -0.083 \text{ kJ g}^{-1}$.

7-36 In every case the applicable equation is:

$$\Delta H^\circ = \sum_{\text{products}} \Delta H_f^\circ \;-\; \sum_{\text{reactants}} \Delta H_f^\circ$$

Enthalpies of formation are always tabulated on a "per amount" basis. The ΔH_f°'s from Appendix D are given per mole, so each must be multiplied by the number of moles appearing in the balanced equation.

(a)

$$\Delta H^\circ = 2 \text{ mol}\left(33.18\frac{\text{kJ}}{\text{mol}}\right) - 1 \text{ mol}\left(0\frac{\text{kJ}}{\text{mol}}\right) - 2 \text{ mol}\left(90.25\frac{\text{kJ}}{\text{mol}}\right) = -114.14 \text{ kJ}$$

(b)

$$\Delta H^\circ = 2 \text{ mol}\left(-110.52\frac{\text{kJ}}{\text{mol}}\right) - 1 \text{ mol}\left(-393.51\frac{\text{kJ}}{\text{mol}}\right) = +172.47 \text{ kJ}$$

(c) We omit the somewhat repetitious listing of the units:

$$\Delta H^\circ = 3(-241.82) + 2(33.18) - 7/2(0) - 2(-46.11) = -566.88 \text{ kJ}$$

(d) $\Delta H^\circ = 1(0) + 1(-110.52) - 1(-241.82) - 1(0) = +131.30 \text{ kJ}$

7-38 (a) The reaction is

$$2\,\text{Al}(s) + \text{Fe}_2\text{O}_3(s) \rightleftharpoons 2\,\text{Fe}(s) + \text{Al}_2\text{O}_3(s)$$

The standard enthalpy change is computed by combining the ΔH_f°'s from Appendix D:

$$\Delta H^\circ = 2 \underbrace{(0)}_{\text{Fe}(s)} + 1\underbrace{(-1675.7)}_{\text{Al}_2\text{O}_3(s)} - 2\underbrace{(0)}_{\text{Al}(s)} - 1\underbrace{(-824.2)}_{\text{Fe}_2\text{O}_3(s)} = -851.5 \text{ kJ}$$

(b) According to the result in part (a), 851.5 kJ of heat is given off at constant pressure when one mole of $Fe_2O_3(s)$ reacts. One mole of Fe_2O_3 is 159.69 g, so the amount of heat given off by the reaction of 3.21 g of $Fe_2O_3(s)$ is

$$q = -851.5 \text{ kJ mol}^{-1} \times \frac{3.21 \text{ g}}{159.69 \text{ g mol}^{-1}} = -17.1 \text{ kJ}.$$

7-40 (a) The reaction is a dissolution:

$$NH_4NO_3(s) \rightleftharpoons NH_4^+(aq) + NO_3^-(aq)$$

We calculate the $\Delta H°$ using the values in Appendix D for the $\Delta H_f°$ of the two ions and the solid salt:

$$\Delta H° - 1(-132.51) + 1(-205.0) - 1(-365.56) = +28.05 \text{ kJ}$$

(b) The 15.0 g of NH_4NO_3 is 0.187 mol of NH_4NO_3. Dissolving this much ammonium nitrate to give the ions in their standard states (at a concentration of 1 M) would have a $\Delta H°$ of 5.26 kJ. In the problem, the final concentrations of the two ions are 2.5 M, not 1 M, so, in principle, a slightly different ΔH is expected. We lack the data to evaluate this difference and simply conclude that the reaction absorbs about 5.3 kJ from its immediate surroundings, which are 100 g of solution. The heat capacity of these surroundings is 418 J K^{-1}, so their temperature change as they give up the 5.3 kJ to the system is -12.6 K. The final temperature of the solution is therefore 7.4°C.

(c) The dissolution of ammonium nitrate in water could be used in cold packs for first aid.

7-42 The combustion reaction is

$$C_6H_{10}(l) + 17/2\,O_2(g) \rightarrow 6\,CO_2(g) + 5\,H_2O(l)$$

for which the standard enthalpy change is -3731.7 kJ.

$$-3731.7 \text{ kJ} = 6(-393.51) + 5(-285.83) - 1(\Delta H_f°(C_6H_{10})(l)) - 17/2(0)$$

Solving for the unknown gives the molar enthalpy of formation of liquid cyclohexene as -58.5 kJ mol^{-1}.

7-44 (a) The combustion of benzoic acid is represented

$$C_6H_5COOH(s) + \frac{15}{2}O_2(g) \rightarrow 7\,CO_2(g) + 3\,H_2O(l)$$

(b)
$$\Delta E = q_V = C\Delta T = (9382 \text{ J K}^{-1}) \times (2.15 \text{ K}) = 2.02 \times 10^4 \text{ J}$$

The heat absorbed by the reaction within the calorimeter is the negative of this. We therefore have for the reaction of benzoic acid:

$$\left(\frac{-2.02 \times 10^4 \text{ J}}{0.800 \text{ g}}\right) \times \left(\frac{122.12 \text{ g}}{1 \text{ mol}}\right) = 3.08 \times 10^6 \text{ J mol}^{-1}$$

The answer is therefore $\Delta E^\circ = -3.08 \times 10^6 \text{ J mol}^{-1} \times 1.000 \text{ mol} = -3.08 \times 10^3 \text{ kJ}$.

(c) Write $\Delta H^\circ = \Delta E^\circ + \Delta n_g RT$ where Δn_g is the change in the number of moles of gas on the two sides of the equation. Substituting gives

$$\Delta H^\circ = -3.08 \times 10^6 \text{ J} + (-\frac{1}{2} \text{ mol})(8.315 \text{ J mol}^{-1} \text{ K}^{-1})(298.15 \text{ K}) = -3.08 \times 10^3 \text{ kJ}$$

(d) The necessary relationship is

$$\Delta H^\circ = \sum_{\text{products}} \Delta H_f^\circ - \sum_{\text{reactants}} \Delta H_f^\circ$$

where the quantity on the left is -3080 kJ and all but one of the terms on the right are known:

$$-3080 \text{ kJ} = 7 \underbrace{(-393.51)}_{\text{CO}_2(g)} + 3 \underbrace{(-285.83)}_{\text{H}_2\text{O}(l)} - 1 \text{ mol} \left(\Delta H_f^\circ(\text{C}_6\text{H}_5\text{COOH})\right)$$

$$\Delta H_f^\circ(\text{C}_6\text{H}_5\text{COOH}) = -532 \text{ kJ mol}^{-1}$$

7-46 The structure of the compound is

$$\text{F}-\underset{\underset{\text{F}}{|}}{\overset{\overset{\text{F}}{|}}{\text{C}}}-\underset{\underset{\text{H}}{|}}{\overset{\overset{\text{Cl}}{|}}{\text{C}}}-\text{Cl}$$

It comes from its component elements in their standard states by

$$2\,\text{C}(s) + 3/2\,\text{F}_2(g) + 1/2\,\text{H}_2(g) + \text{Cl}_2(g) \rightarrow \text{C}_2\text{F}_3\text{HCl}_2(g)$$

In the following equation, each tabular value is multiplied by the number of moles of bonds formed (or number of moles of gaseous atoms formed by atomization).

$$\Delta H^\circ = 2 \underbrace{(716.7) \text{ kJ}}_{\text{C}(g)} + 3 \underbrace{(79.0) \text{ kJ}}_{\text{F}(g)} + 1 \underbrace{(218.0) \text{ kJ}}_{\text{H}(g)} + 2 \underbrace{(121.7) \text{ kJ}}_{\text{Cl}(g)} - 3 \underbrace{(441) \text{ kJ}}_{\text{C}-\text{F}}$$

$$-1 \underbrace{(413) \text{ kJ}}_{\text{C}-\text{H}} -1 \underbrace{(348) \text{ kJ}}_{\text{C}-\text{C}} -2 \underbrace{(328) \text{ kJ}}_{\text{C}-\text{Cl}} = -608 \text{ kJ}$$

The molar ΔH_f° of the compound is approximately -608 kJ mol^{-1}.

7-48 The reaction forming one mole of ethane from ethylene and hydrogen in their standard states can be imagined as breaking all the bonds in the reactants, which are one mole of gaseous ethylene and one mole of gaseous hydrogen, and then constructing one mole of ethane from the resulting gaseous atoms.

$$\Delta H^\circ = 4 \underbrace{(413) \text{ kJ}}_{\text{C}-\text{H}} +1 \underbrace{(615) \text{ kJ}}_{\text{C}=\text{C}} +1 \underbrace{(436) \text{ kJ}}_{\text{H}-\text{H}} -6 \underbrace{(413) \text{ kJ}}_{\text{C}-\text{H}}$$

$$-1 \underbrace{(348) \text{ kJ}}_{\text{C}-\text{C}} = -123 \text{ kJ}$$

7-50 (a) On the left side of the equation, there are 6 mol of Hg—Cl and 8 mol Al—Cl bonds; on the right side of the equation there are also 6 mol of Hg—Cl and 8 mol of Al—Cl bonds (note that 2 mol of compound forms). The bond enthalpy of a single bond between two elements is nearly constant from compound to compound, so we expect ΔH to be close to zero.

(b) The proposed structure of Hg_2Cl_4 contains 4 Hg—Cl and 1 Hg—Hg bond. A larger ΔH would be expected, because different bonds are broken than are formed.

7-52

$$n = \frac{54.0 \text{ g}}{39.948 \text{ g mol}^{-1}} = 1.352 \text{ mol}$$

$$\Delta E = 0 \text{ (isothermal, reversible)}$$

$$w = -nRT \ln \frac{V_2}{V_1} = -nRT \ln \frac{P_1}{P_2}$$

$$= -(1.352 \text{ mol})(8.315 \text{ J mol}^{-1} \text{ K}^{-1})(400 \text{ K}) \ln \frac{1.50 \text{ atm}}{4.00 \text{ atm}} = +4410 \text{ J}$$

$$q = -w = -4410 \text{ J}$$

$$\Delta H = 0 \text{ because } \Delta T = 0$$

7-54 For an adiabatic process, $q = 0$, and

$$P_1 V_1^\gamma = P_2 V_2^\gamma$$

Here we have $P_1 = 1.00$ atm, $P_2 = 2.00$ atm, and

$$V_1 = \frac{nRT_1}{P_1} = \frac{(2.00 \text{ mol})(0.08206 \text{ L atm mol}^{-1}\text{K}^{-1})(273.15 \text{ K})}{1.00 \text{ atm}} = 44.83 \text{ L}$$

$$\gamma = \frac{c_P}{c_V} = \frac{c_P}{c_P - R} = \frac{29.3}{29.3 - 8.315} = 1.396$$

$$\left(\frac{V_2}{V_1}\right)^\gamma = \frac{P_1}{P_2} = 0.500$$

$$\frac{V_2}{V_1} = (0.500)^{1/\gamma} = 0.6087$$

$$V_2 = 0.6087(44.83 \text{ L}) = 27.29 \text{ L}$$

$$T_2 = \frac{P_2 V_2}{nR} = \frac{(2.00 \text{ atm})(27.29 \text{ L})}{(2.00 \text{ mol})(0.08206 \text{ L atm mol}^{-1}\text{K}^{-1})} = 332.5 \text{ K}$$

$$\Delta E = nc_V \Delta T = (2.00 \text{ mol})(29.3 - 8.315 \text{ J mol}^{-1} \text{ K}^{-1})(332.5 - 273.15 \text{ K})$$

$$= +2.49 \times 10^3 \text{ J}$$

$$w = \Delta E - q = \Delta E = 2.49 \times 10^3 \text{ J}$$

7-56 The specific heat of a substance changes with the temperature, so that the heat absorbed per gram or per mole of a substance in a temperature change at constant pressure is really an integral

$$q_P = \int_{T_1}^{T_2} c_P dT$$

The dependence of c_P on temperature is usually slight, but it is large in this problem. We estimate the integral as the area under the curve in a plot of c_s versus T. The units of this area are J g^{-1}, which are units of specific heat. The area is arrived at by adding up the areas of a set of columns of width 0.05 K and height equal to the average value of c_s at the two ends of each 0.05 K temperature range. Thus:

$$0.05\left(\frac{2.81 + 3.26}{2} + \frac{3.26 + 3.79}{2} + \frac{3.79 + 4.42}{2} + \frac{4.42 + 5.18}{2}\right.$$

$$\left. + \frac{5.18 + 6.16}{2} + \frac{6.16 + 7.51}{2} + \frac{7.51 + 9.35}{2}\right) = 1.82 \text{ J g}^{-1}$$

7-58

$$q = \Delta H = (38 \text{ J mol}^{-1} \text{ K}^{-1})(0 - (-30) \text{ K})(1.00 \text{ mol})$$
$$+(1.00 \text{ mol})(6007 \text{ J mol}^{-1} \text{ K}^{-1})$$
$$+(75 \text{ J mol}^{-1} \text{ K}^{-1})(100 - 0 \text{ K})(1.00 \text{ mol})$$
$$+(1.00 \text{ mol})(40660 \text{ J mol}^{-1})$$
$$+(36 \text{ J mol}^{-1} \text{ K}^{-1})(140 - 100 \text{ K})(1.00 \text{ mol})$$
$$= 5.675 \times 10^4 \text{ J} = 56.7 \text{ kJ}$$

$\Delta E = \Delta H - \Delta(PV) = \Delta H - (P_{\text{final}}V_{\text{final}} - P_{\text{initial}}V_{\text{initial}}) \approx \Delta H - P_{\text{final}}V_{\text{final}}$ because the volume of ice is much smaller than that of steam.

$$PV = nRT = (1.00 \text{ mol})(8.315 \text{ J mol}^{-1} \text{ K}^{-1})(413 \text{ K}) = 3.43 \times 10^3 \text{ J}$$

$$\Delta E = 5.675 \times 10^4 - 3.43 \times 10^3 = 5.33 \times 10^4 \text{ J} = 53.3 \text{ kJ}$$

$$w = \Delta E - q = 53.3 - 56.7 = -3.4 \text{ kJ}$$

7-60 (a) Substituting in $PV = nRT$ gives $P = 1.000$ atm.

(b) $q = +3.000$ kJ.

(c) The strong rigid container neither expands nor contracts during the heating. There is therefore no pressure-volume work. Because there is no other mechanical linkage to the surroundings, $w = 0$.

(d) The temperature change of the gas is given by

$$\Delta T = \frac{q}{n\, c_V} = \frac{3000 \text{ J}}{(1.000 \text{ mol})(12.47 \text{ J mol}^{-1}\text{K}^{-1})} = 240.6 \text{ K}$$

so the final T is 513.7 K .

(e) We use the ideal gas equation: $P = 1.881$ atm.

(f) $\Delta E = q + w = 3000 + 0 = +3000$ J.

(g) $\Delta H = \Delta E + \Delta(PV)$. We know the pressure and volume before the heating and we also know them both after the heating. Therefore we can compute $\Delta(PV)$. It is 19.73 L atm, which is 2000 J. Hence, $\Delta H = 3000 + 2000 = +5000$ J

(h) The law of conservation of energy applies to energy, not enthalpy. The change in enthalpy is equal to the heat absorbed only if the pressure is held constant.

7-62 The energy released in the combustion of gasoline in fueling the car to travel a kilometer is

$$1 \text{ km} \times \left(\frac{1 \text{ L gasoline}}{8.0 \text{ km}}\right) \times \left(\frac{0.68 \text{ kg}}{1 \text{ L}}\right) \times \left(\frac{48 \times 10^3 \text{ kJ}}{1 \text{ kg}}\right) = 4080 \text{ kJ}$$

The energy released in the oxidation of food is about $(100 / 0.3)$ kJ or 330 kJ. The energy savings is around 3750 kJ.

7-64

$$q_{\text{water}} = (50.0 \text{ g})(4.184 \text{ J K}^{-1} \text{ g}^{-1})(23.36 - 25.00 \text{ K}) = -343 \text{ J}$$

$$\Delta H^\circ = \left(\frac{+343 \text{ J}}{1 \text{ g KClO}_3}\right)\left(\frac{122.55 \text{ g KClO}_3}{1 \text{ mol KClO}_3}\right) = +4.20 \times 10^4 \text{ J} = +42.0 \text{ kJ}$$

7-66 The ΔH° of the reaction is the difference between the ΔH_f°'s of the products and the reactants. It is

$$2 \underbrace{(96.23)}_{\text{HgBr}(g)} -1 \underbrace{(-206.77)}_{\text{Hg}_2\text{Br}_2(s)} = +399.2 \text{ kJ}$$

7-68

$$\begin{aligned} \Delta H_1 &= (1 \text{ mol})(39.9 \text{ J mol}^{-1} \text{ K}^{-1})(298 - 500 \text{ K}) \\ &+ (\tfrac{1}{2} \text{ mol})(29.4 \text{ J mol}^{-1} \text{ K}^{-1})(298 - 500 \text{ K}) \\ &= -1.103 \times 10^4 \text{ J} \end{aligned}$$

$$\Delta H_2 = \Delta H^\circ = -395.72 - (-296.83) = -98.89 \text{ kJ}$$

$$\Delta H_3 = (1 \text{ mol})(50.7 \text{ J mol}^{-1} \text{ K}^{-1})(500 - 298 \text{ K}) = +1.024 \times 10^4 \text{ J}$$

$$\Delta H = \Delta H_1 + \Delta H_2 + \Delta H_3 = -98.89 - 11.03 + 10.24 \text{ kJ} = -99.7 \text{ kJ}$$

7-70

$$\frac{46.0 \text{ g}}{32.00 \text{ g mol}^{-1}} = 1.4375 \text{ mol O}_2$$

$$V_1 = \frac{(1.4375 \text{ mol})(0.08206 \text{ J mol}^{-1} \text{ K}^{-1})(400 \text{ K})}{1.00 \text{ atm}} = 47.2 \text{ L}$$

$$\gamma = \frac{c_P}{c_V} = \frac{29.4}{29.4 - 8.315} = 1.394$$

$$\left(\frac{V_2}{V_1}\right)^\gamma = \frac{P_1}{P_2} = \frac{1.00 \text{ atm}}{0.60 \text{ atm}} = 1.667$$

$$\frac{V_2}{V_1} = (1.667)^{1/\gamma} = 1.442$$

$$V_2 = 1.442\, V_1 = 68.1 \text{ L}$$

$$T_2 = \frac{(0.60 \text{ atm})(68.1\text{L})}{(1.4375 \text{ mol})(0.08206 \text{ L atm mol}^{-1}\text{K}^{-1})} = 346 \text{ K}$$

$$\Delta E_1 = nc_V \Delta T = (1.4375 \text{ mol})(29.4{-}8.315 \text{ J mol}^{-1} \text{ K}^{-1})(346{-}400 \text{ K}) = -1630 \text{ J}$$

$$w_1 = \Delta E_1 = -1630 \text{ J}$$

In the second step, V decreases at constant T from 68.1 to 47.2 L,

$$P_{\text{final}} = \frac{68.1 \text{ L}}{47.2 \text{ L}} \times 0.60 \text{ atm} = 0.866 \text{ atm}$$

$$w_2 = -nRT \ln \left(\frac{47.2 \text{ L}}{68.1 \text{ L}}\right) = +1.52 \times 10^3 \text{ J} = -q_2$$

$$\Delta E_2 = 0 \quad \text{(isothermal)}$$

Thus the overall change is

$$\Delta E = -1630 \text{ J}$$

$$w = -1630 + 1520 = -110 \text{ J}$$

$$q = -1520 \text{ J}$$

7-72 The reaction is

$$ClF_3(g) + 4 \text{ Li}(s) \rightarrow LiCl(s) + 3 \text{ LiF}(s)$$

and the amounts are 0.3472 mol ClF_3 and 2.492 mol Li. The limiting reactant is ClF_3. If one mole of ClF_3 reacted,

$$\Delta H^\circ = -408.61 + 3(-615.97) - (-163.2) = -2093.3 \text{ kJ}$$

The ΔH for 0.3472 mol reacting is smaller by a factor of 0.3472, giving $\Delta H = -727$ kJ. The amount of heat <u>evolved</u> is 727 kJ.

7-74 (a) The point of drawing all the Lewis structures is to identify the type and number of bonds in the compounds in parts (b) through (d). The O_2 has an O=O double bond; CO_2 has two C=O double bonds; H—O—H has two O—H single bonds; CH_4 has four C—H single bonds, C_2H_5OH has five C—H single bonds, one C—C single bond, one C—O single bond, and one O—H single bond; C_8H_{18} has 18 C—H single bonds and seven C—C single bonds,

(b) For the burning of one mole of methane we have:

$$\Delta H^\circ = -\underbrace{4(463)}_{4 \text{ O–H}} - \underbrace{2(728)}_{2 \text{ C=O}} + \underbrace{4(413)}_{4 \text{ C–H}} + \underbrace{2(498)}_{2 \text{ O=O}} = -660 \text{ kJ}$$

(c) For the burning of one mole of octane the equation is:

$$\Delta H^\circ = -\underbrace{18(463)}_{18 \text{ O–H}} - \underbrace{16(728)}_{16 \text{ C=O}} + \underbrace{18(413)}_{18 \text{ C–H}} + \underbrace{7(348)}_{7 \text{ C–C}} + \underbrace{25/2(498)}_{25/2 \text{ O=O}} = -3887 \text{ kJ}$$

(d) For the burning of one mole of ethanol the equation is:

$$\Delta H^\circ = -\underbrace{6(463)}_{6 \text{ O–H}} - \underbrace{4(728)}_{4 \text{ C=O}} + \underbrace{5(413)}_{5 \text{ C–H}} + \underbrace{1(348)}_{1 \text{ C–C}} + \underbrace{1(351)}_{1 \text{ C–O}} + \underbrace{1(463)}_{1 \text{ O–H}} + \underbrace{3(498)}_{3 \text{ O=O}} = -969 \text{ kJ}$$

7-76 (a) Before the expansion, the average velocity is zero, and the average speed is 1455 m s^{-1}.

(b)

$$\frac{T_2}{T_1} = \left(\frac{V_1}{V_2}\right)^{\gamma-1} = \left(\frac{V_1^\gamma}{V_2^\gamma}\right)^{(\gamma-1)/\gamma} = \left(\frac{P_2}{P_1}\right)^{(\gamma-1)/\gamma}$$

$$\gamma = \frac{c_P}{c_V} = \frac{5/2R}{3/2R} = \frac{5}{3} = 1.667$$

$$\frac{\gamma-1}{\gamma} = 0.40$$

$$T_2 = T_1 \left(\frac{P_2}{P_1}\right)^{0.40} = 400 \text{ K} \left(\frac{1.0 \text{ atm}}{50 \text{ atm}}\right)^{0.40} = 84 \text{ K}$$

(c) Change in random thermal energy per molecule is

$$(3/2)k_B T_f - (3/2)k_B T_i = (3/2)(1.381 \times 10^{-23} \text{ J K}^{-1}) \times (84 - 400 \text{ K}) = -6.55 \times 10^{-21} \text{ J}$$

This energy must go into net directed motion $1/2\, m\bar{v}^2$

$$1/2\, m\bar{v}^2 = 6.55 \times 10^{-21} \text{ J}$$

$$\bar{v}^2 = \frac{2(6.55 \times 10^{-21} \text{ J})(6.022 \times 10^{23} \text{ mol}^{-1})}{(4.0026 \times 10^{-3} \text{ kg mol}^{-1})} = 1.97 \times 10^6 \text{ m}^2 \text{ s}^{-2}$$

$$\bar{v} = 1400 \text{ m s}^{-1}$$

Chapter 8

Spontaneous Processes and Thermodynamic Equilibrium

8-2 All five processes occur in the real world and thus are spontaneous.

    (a) System: HCl and NaOH solutions.

        Surroundings: Container, buret, air.

        Constraint removed: Opening the stopcock of the buret removes the initial constraint that the two solutions are separated.

    (b) System: $Zn(s)$ and $HCl(aq)$.

        Surroundings: Container, air.

        Constraint removed: The imaginary barrier between the zinc and the HCl solution is removed, allowing zinc ions and gaseous hydrogen to form.

    (c) System: Rubber band.

        Surroundings: Weight, air.

        Constraint removed: The length of the rubber band becomes unconstrained.

    (d) System: Gas

        Surroundings: Chamber, piston.

        Constraint removed: The volume of the gas in the chamber becomes unconstrained.

    (e) System: Water (ice).

        Surroundings: Refrigerator.

        Constraint removed: Thermal barrier separating the water in the tray from its surroundings.

8-4 Label the three equal volumes L (left), C (center), and R (right). Each of the four molecules could be in any one of the three and thus has three equally probable ways of going into the overall volume. The number of microstates is therefore 3^4 or 81.

(b) In only one of the microstates are all four molecules in L. Thus the probability is 1/81.

8-6 A system containing the "chemically mixed" boron halides has greater entropy than a system of just BCl_3 and BF_3. It has the same number of gas-phase molecules, but more distinguishable kinds of molecules, hence more microstates and higher entropy.

8-8 Probability that all N_2 is in left half $= \left(\frac{1}{2}\right)^{2N_o}$.

Probability that all O_2 is in right half $= \left(\frac{1}{2}\right)^{N_o}$.

Joint probability that both are true $= \left(\frac{1}{2}\right)^{2N_o} \times \left(\frac{1}{2}\right)^{N_o} = \left(\frac{1}{2}\right)^{3N_o} = \left(\frac{1}{2}\right)^{1.81\times10^{24}} = \dfrac{1}{10^{5.4\times10^{23}}}$

8-10 (a) Computer constructed: $\Delta S < 0$.   (b) Gas leaks out: $\Delta S > 0$.   (c) Solid sublimes: $\Delta S > 0$.

8-12 (a)
$$\epsilon = \frac{T_h - T_l}{T_h} = \frac{400 - 300}{400} = 0.250$$

(b)
$$\epsilon = \frac{q_h + q_l}{q_h} = \frac{1000 + q_l}{1000} = 0.250$$
$$q_l = 250 - 1000 = -750 \text{ J}$$

(c)
$$\epsilon = \frac{-w_{\text{net}}}{q_h} = \frac{-w_{\text{net}}}{1000 \text{ J}} = 0.250$$
$$-w_{\text{net}} = 250 \text{ J} = \text{ maximum work performed } by \text{ engine}$$

8-14 The melting point of tetraphenylgermane is 505.65 K. The molar enthalpy of fusion is
$$381.03 \text{ g mol}^{-1} \times 106.7 \text{ J g}^{-1} = 4.066 \times 10^4 \text{ J mol}^{-1}$$

We combine these numbers to get the molar entropy of fusion as follows:
$$\Delta S_{\text{fus}} = \frac{\Delta H_{\text{fus}}}{T_{\text{fus}}} = \frac{4.066 \times 10^4 \text{ J mol}^{-1}}{505.65 \text{ K}} = +80.40 \text{ J K}^{-1}\text{mol}^{-1}$$

8-16 Trouton's rule states that the molar entropy of vaporization of most substances is near 88 J K^{-1}mol^{-1}. Gas and liquid forms of a substance are in equilibrium at the normal boiling point, so we can write:

$$T_b = \frac{\Delta H_{vap}}{\Delta S_{vap}} = \frac{16.15 \times 10^3 \text{ J mol}^{-1}}{88 \text{ J K}^{-1}\text{mol}^{-1}} = 184 \text{ K}$$

8-18 Let us designate the reversible isochoric heating process by subscript (1) and the isothermal reversible expansion by subscript (2).

$$n = \frac{60.0 \text{ g}}{80.91 \text{ g mol}^{-1}} = 0.7415 \text{ mol}$$

$$w_1 = 0 \quad \text{(process is at constant volume)}$$

$$c_V = c_P - R = 29.1 - 8.3 = 20.8 \text{ J mol}^{-1} \text{ K}^{-1}$$

$$q_1 = \Delta E = nc_V \Delta T = (0.7415 \text{ mol})(20.8 \text{ J mol}^{-1} \text{ K}^{-1})(500 - 300 \text{ K}) = 3,085 \text{ J}$$

$$\Delta H_1 = nc_P \Delta T = (0.7415)(29.1)(500 - 300) = 4,316 \text{ J}$$

$$\Delta S_1 = nc_V \ln T_2/T_1 = (0.7415)(20.8) \ln(500/300) = 7.88 \text{ J K}^{-1}$$

$$\Delta H_2 = \Delta E_2 = 0 \quad \text{(isothermal process on ideal gas)}$$

At constant volume,
$$\frac{P_2}{P_1} = \frac{T_2}{T_1} = \frac{500 \text{ K}}{300 \text{ K}} = 1.667$$

For an isothermal expansion from P_i to P_f we have $V_f/V_i = P_i/P_f$. Here the initial pressure P_i is P_2 and the final pressure P_f is P_1, so that

$$q_2 = -w_2 = nRT_2 \ln(P_2/P_1)$$

$$= (0.7415 \text{ mol})(8.315 \text{ J mol}^{-1} \text{ K}^{-1})(500 \text{ K}) \ln 1.667 = 1575 \text{ J}$$

$$\Delta S_2 = \frac{q_2}{T_2} = \frac{1575 \text{ J}}{500 \text{ K}} = 3.15 \text{ J K}^{-1}$$

$$\Delta E = \Delta E_1 + \Delta E_2 = 3,085 \text{ J} = 3.08 \text{ kJ}$$

$$q = q_1 + q_2 = 4,660 \text{ J} = 4.66 \text{ kJ}$$

$$w = w_1 + w_2 = -1,575 \text{ J} = -1.58 \text{ kJ}$$

$$\Delta H = \Delta H_1 + \Delta H_2 = 4,316 \text{ J} = 4.32 \text{ kJ}$$

$$\Delta S = \Delta S_1 + \Delta S_2 = 11.0 \text{ J K}^{-1}$$

8-20 Consider first the *reversible* heating of 1.00 mol water from 25°C to 150°C. This process consists of three steps: heating water to 100°C, evaporating it, and heating steam to 150°C.

$$
\begin{aligned}
q &= (1.00 \text{ mol})(75.4 \text{ J mol}^{-1} \text{ K}^{-1})(100 - 25 \text{ K}) \\
&+ (1.00 \text{ mol})(40680 \text{ J mol}^{-1} \text{ K}^{-1}) \\
&+ (1.00 \text{ mol})(36.0 \text{ J mol}^{-1} \text{ K}^{-1})(150 - 100 \text{ K}) \\
&= 48,135 \text{ J}
\end{aligned}
$$

$$
\Delta S = n c_P(liq) \ln \frac{T_b}{T_i} + n \frac{\Delta H_{vap}}{T_b} + n c_P(gas) \ln \frac{T_f}{T_b}
$$

$$
\begin{aligned}
&= (1.00 \text{ mol})(75.4 \text{ J mol}^{-1} \text{ K}^{-1}) \times \ln(373.15/298.15) \\
&+ (1.00 \text{ mol}) \frac{40680 \text{ J mol}^{-1}}{373.15 \text{ K}} \\
&+ (1.00 \text{ mol})(36.0 \text{ J mol}^{-1} \text{ K}^{-1}) \times \ln(423.15/373.15)
\end{aligned}
$$

$$
= 130.5 \text{ J K}^{-1}
$$

Now consider the irreversible process in the flash evaporation. Because S is a function of state, ΔS for the water is unchanged.

$$
\Delta S_{water} = 130.5 \text{ J K}^{-1}
$$

From the point of view of the *iron*, however, the heat $q = 48,135$ J is removed at a constant temperature of 150°C = 423.15 K. Thus the final state of the iron is the same as would be achieved by removing 48,135 J reversibly at fixed temperature.

$$
\Delta S_{iron} = \frac{q}{T} = -\frac{48,135 \text{ J}}{423.15 \text{ K}} = -113.8 \text{ J K}^{-1}
$$

$$
\Delta S_{total} = 130.5 - 113.8 = 16.7 \text{ J K}^{-1} > 0
$$

8-22 (a)

$$
\Delta H = n c_P \Delta T = (1.00 \text{ mol})(25.1 \text{ J mol}^{-1} \text{ K}^{-1})(273.15 - 373.15 \text{ K}) = -2510 \text{ J}
$$

$$
\Delta S = n c_P \ln \frac{T_f}{T_i} = (1.00 \text{ mol})(25.1 \text{ J mol}^{-1} \text{ K}^{-1}) \ln \frac{273.15}{373.15} = -7.83 \text{ J K}^{-1}
$$

(b) $\Delta S_{iron} = -7.83$ J K^{-1} because S is a function of state and the initial and final states of the iron are the same as in part (a). For the water, heat is being added at a constant temperature of 273.15 K. Its final state is the same as would be found if $q_p = 2510$ J were added at $T = 273.15$ K.

$$
\Delta S_{water} = \frac{2510 \text{ J}}{273.15 \text{ K}} = +9.19 \text{ J K}^{-1}
$$

$$
\Delta S_{total} = 1.36 \text{ J K}^{-1}
$$

8-24 (a) The $\Delta S°$ of the reaction as written is:

$$1\underbrace{(130.57)}_{H_2(g)}+1\underbrace{(213.63)}_{CO_2(g)}+1\underbrace{(243.30)}_{CH_3NH_2(g)}-1\underbrace{(282.4)}_{CH_3COOH(g)}-1\underbrace{(192.34)}_{NH_3(g)}=+112.8 \text{ J K}^{-1}$$

where the numbers in parenthesis are standard molar entropies (with units of J K^{-1}mol^{-1}) and the other numbers are the numbers of moles from the balanced equation.

(b) It is a certainty that the suggested new reactants, which comprise a mole of a solid and a mole of a liquid, have less entropy than the two moles of gaseous reactants in the equation in part (a). The products have the same entropy so the change in entropy is larger than in part (a).

8-26 We compute the standard entropy changes ($\Delta S°$'s) of these reactions by adding 1 mol times the tabulated molar $S°$ of the gaseous halogen to 1 mol times the tabulated molar $S°$ of gaseous hydrogen and then subtracting the sum from 2 mol times the molar $S°$ of the hydrogen halide. The answers are 14.10, 20.07, 21.26 and 21.81 J K^{-1}, moving down the periodic table from fluorine to iodine. The entropy change gets more positive going down the table; the $\Delta S°$ for the reaction forming HF is less than extrapolation from the values for the other reactions would predict.

8-28 The fundamental criterion for the spontaneity in the decomposition of quartz or any other process is whether ΔS_{univ} is positive or negative in the process. The large positive entropy change quoted for the decomposition of quartz is for a specific system consisting of the reactants and products only. The ΔS of the surroundings must be even larger and negative in sign in the proposed process to prevent the process from being spontaneous.

8-30 (a) The solvent, water, becomes ordered around the dissolved ions, which reduces its entropy. This must more than compensate for the increased entropy of the ions themselves.

(b)
$$\Delta S° = -53.1 + 2(-13.8) - 68.87 = -149.6 \text{ J K}^{-1}$$

Fluoride ion forms hydrogen bonds with water, which lowers the entropy of the products still further.

8-32 (a) At constant temperature and pressure we have the relationship $\Delta G = \Delta H - T\Delta S$. Substitution then gives

$$\Delta G = -2.1 \times 10^3 \text{ J} - (243.15 \text{ K})(-7.4 \text{ J K}^{-1}) = -300 \text{ J}$$

(b) If the amount of tin is increased from 1.00 mol to 2.50 mol, then the ΔG increases by the same factor of 2.50, from -300 to -750 J, because of the larger size of the system.

(c) The ΔG of the process at $-30°C$ is negative, so tin tends to convert spontaneously from white tin to gray tin at this low temperature.

(d) At equilibrium at constant temperature and pressure, $\Delta G = 0$. We compute the temperature that makes this condition happen:

$$T = \frac{\Delta H}{\Delta S} = \frac{-2.1 \times 10^3 \text{ J}}{-7.4 \text{ J K}^{-1}} = 284 \text{ K}$$

This assumes that ΔH and ΔS are independent of temperature.

8-34 Consider three steps: (1) cooling superheated solid from 25°C to 0°C, (2) letting it melt at 0°C; and (3) heating the liquid to 25°C. Then

$$\Delta H = \Delta H_1 + \Delta H_2 + \Delta H_3$$

$$= nc_P(solid)(0 - 25) + 6007 \text{ J} + nc_P(liquid)(25 - 0) = 6007 \text{ J}$$

because $c_P(solid) \approx c_P(liquid)$. For the same reason,

$$\Delta S \approx \Delta S_2 = \frac{6007 \text{ J}}{273.15 \text{ K}} = 21.99 \text{ J K}^{-1}$$

$$\Delta G = \Delta H - T\Delta S = 6007 \text{ J} - (21.99 \text{ J K}^{-1})(298.15 \text{ K}) = -550 \text{ J}$$

8-36 Add the second reaction to 38 times the first, and combine the free energy changes as well, giving

$$\Delta G_{tot} = -2872 + 38(34.5) = -1561 \text{ kJ}$$

Because $\Delta G < 0$, this reaction is spontaneous. The fraction of free energy stored is

$$\frac{38(34.5 \text{ kJ})}{2872 \text{ kJ}} = 0.456$$

8-38 (a) The reaction written in the problem for the preparation of phosgene has $\Delta H° = -108.3$ kJ, and $\Delta S° = -136.99$ J K^{-1}. The quotient $\Delta H°/\Delta S°$ determines the temperature at which $\Delta G = 0$; it is 790 K. Below this temperature the reaction is spontaneous. Because of the imperfection of the assumption that ΔH and ΔS are temperature-independent, there is little point in using more that two significant figure in the answer.

(b) For the reaction as written in the problem, $\Delta H° = -78.04$ kJ, and $\Delta S° = 494.07$ J K^{-1}. The reaction is spontaneous at all temperatures.

(c) We use $\Delta H_f°$ and $S°$ values tabulated for graphite (not diamond!) and for wüstite. Then, for the reaction with the coefficients given in the problem, $\Delta H° = 155.75$ kJ, and $\Delta S° = 161.61$ J K^{-1}. The reaction is spontaneous above about 960 K.

8-40 (a) The $\Delta G°$ of the reaction is

$$\Delta G° = 2 \underbrace{(0)}_{W(s)} +3 \underbrace{(-394.36)}_{CO_2(g)} -3 \underbrace{(0)}_{C(s)} -2 \underbrace{(-764.08)}_{WO_3(s)} = +345.08 \text{ kJ}$$

The process is *not* spontaneous at room temperature.

(b) Without doing any calculations, the fact that a gas is generated from solids means that $\Delta S°$ of this reaction is positive. It follows that a high enough temperature will make ΔG negative no matter what the value of $\Delta H°$.

8-42 The container is insulated so no heat can pass into it from the surroundings or out of it to the surroundings: $q = 0$. Further, the container is rigid so that it cannot expand or contract during the process. There is no other mechanical linkage or electrical linkage to the surroundings. Hence $w = 0$. By the first law then, $\Delta E = 0$. The total energy of the contents of the container remains the same during the process.

We know from experience that at equilibrium the hot water will have cooled and that some or all of the ice will have melted spontaneously. The change occurs in isolation from the surroundings so $\Delta S_{surr} = 0$. The second law states that for this spontaneous process $\Delta S_{univ} > 0$. Hence $\Delta S_{system} > 0$.

8-44 The easy way to solve this is to remember that S is a function of state, so the easiest way to calculate ΔS is to consider a different path between the same two states: a constant pressure heating.

$$\Delta S = nc_P \ln \frac{T_2}{T_1} = (1.00 \text{ mol}) \times \frac{5}{2}(8.315 \text{ J mol}^{-1} \text{ K}^{-1}) \ln \frac{400}{300} = 5.98 \text{ J K}^{-1}$$

The same answer is found, with much more work, by adding ΔS for the four steps given.

8-46 Although the process is carried out in an insulated vessel so that there is no heat exchange with the environment and consequently no entropy change for the environment, the system increases in entropy because the process is irreversible. Let us find a two-step reversible process leading to the same final state.

Step 1 : 54.0 g $H_2O(l)$ at 0°C $\rightarrow$ 54.0 g $H_2O(l)$ at t_f.

Step 2: 27.0 g $H_2O(l)$ at 100°C $\rightarrow$ 27.0 g $H_2O(l)$ at t_f.

Because $q_{total} = 0$,

$$54.0 \text{ g} \times 4.18 \text{ J K}^{-1} \text{ g}^{-1}(t_f - 0) + 27.0 \text{ g} \times 4.18 \text{ J K}^{-1}\text{g}^{-1}(t_f - 100) = 0$$

$$t_f = 33.33°C = 306.48 \text{ K}$$

$$\Delta S_1 = n_1 c_V \ln \frac{306.48}{273.15}$$

$$= \frac{54.0 \text{ g}}{18.016 \text{ g mol}^{-1}} \times (4.18 \text{ J K}^{-1} \text{ g}^{-1} \times 18.016 \text{ g mol}^{-1}) \ln \frac{306.48}{273.15} = 25.99 \text{ J K}^{-1}$$

$$\Delta S_2 = n_2 c_V \ln \frac{306.48}{373.15} = -22.21 \text{ J K}^{-1}$$

$$\Delta S_{sys} = \Delta S_1 + \Delta S_2 = 3.8 \text{ J K}^{-1}$$

8-48 The chance that the first of the 500 atoms is in the left half of the optical trap is 1/2. We assume that the gas of sodium atoms behaves ideally so that the chance that a second sodium atom is in the left half of the optical trap is not influenced by the presence of the first. It is also 1/2. The chance that both by these two atoms are the left half is $1/2 \times 1/2 = 1/4$. By extension, p, the chance that all 500 sodium atoms are in the left half, is

$$p = \left(\frac{1}{2}\right)^{500}$$

To compute p we take the logarithm of both sides of this equation

$$\log p = 500 \log \frac{1}{2} = 500 \times (-0.30103) = -150.51$$

and then the antilog of both sides

$$p = 10^{-150.51} = 10^{0.49} \times 10^{-151} = 3 \times 10^{-151}$$

Even with this very small sample of gas, the chance of "spontaneous congregation" on one side of the vessel is essentially zero.

8-50 (a) 2^{N_o} microstates $= 10^{1.81 \times 10^{23}} = \Omega$

(b) $\Delta S = k_B \ln \Omega = (1.38 \times 10^{-23} \text{ J K}^{-1})(1.81 \times 10^{23} \ln 10) = 5.76 \text{ J K}^{-1}$

8-52 (a) The process takes 2.00 mol of an ideal monatomic gas from $V = 19.15$ L to 38.15 L and from $T = 350$ K to 272 K. To calculate ΔS_{sys} we connect the same two states by a reversible path: first expanding reversibly at constant temperature

$$\Delta S_1 = nR \ln V_2/V_1 = (2.00 \text{ mol})(8.315 \text{ J mol}^{-1} \text{ K}^{-1} \ln 2) = 11.5 \text{ J K}^{-1}$$

and then cooling at constant volume:

$$\Delta S_2 = nC_V \ln T_2/T_1$$

$$= (2.00 \text{ mol})(3/2)(8.315 \text{ J mol}^{-1} \text{ K}^{-1}) \ln(272/350) = -6.3 \text{ J K}^{-1}$$

Thus $\Delta S_{sys} = \Delta S_1 + \Delta S_2 = 5.2 \text{ J K}^{-1}$

Because $q = 0$, $\Delta S_{surr} = 0$, and $\Delta S_{univ} = 5.2 \text{ J K}^{-1}$

(b)

$$\begin{aligned}
\Delta G &= \Delta H - \Delta(TS) = nc_P \Delta T - (T_f S_f - T_i S_i) \\
&= (2.00 \text{ mol})(5/2)(8.315 \text{ J mol}^{-1} \text{ K}^{-1})(272 - 350 \text{ K}) \\
&\quad -(272 \text{ K})(158.2 + 5.2 \text{ J K}^{-1}) + (350 \text{ K})(158.2 \text{ J K}^{-1}) \\
&= -3243 \text{ J} - 44,445 \text{ J} + 55,370 \text{ J} = +7680 \text{ J}
\end{aligned}$$

Note that $\Delta G > 0$ in this process even though it is spontaneous. $\Delta G < 0$ is a criterion for spontaneity only at constant temperature and pressure. $\Delta S_{univ} > 0$ is a more general criterion.

8-54 At equilibrium between solid and gaseous iodine:

$$\Delta G = 0 = \Delta H - T\Delta S$$

so that $T = \Delta H/\Delta S$. The ΔH and ΔS of the process are well approximated by its $\Delta H°$ and $\Delta S°$. Compute these for the reaction $I_2(s) \rightleftharpoons I_2(g)$ in the usual way, using the data in Appendix D. Then:

$$T = \frac{62.44 \times 10^3 \text{ J}}{144.44 \text{ J K}^{-1}} = 432.3 \text{ K}$$

The computation indicates that solid and vapor are in equilibrium at 159.1°C when the pressure is 1 atm. The melting point of $I_2(s)$ is 113.5° at $P = 1$ atm (from Appendix F). The solid-gas equilibrium at 1 atm cannot be accomplished because the solid melts first.

8-56 We have data on the enthalpies and free energies of formation of the two isomers of $Pt(NH_3)_2I_2$. This means that we know $\Delta G°$ and $\Delta H°$ of the reactions

$$Pt(s) + N_2(g) + 3H_2(g) + I_2(s) \rightleftharpoons cisPt(NH_3)_2I_2(s)$$

$$\text{Pt}(s) + \text{N}_2(g) + 3\,\text{H}_2(g) + \text{I}_2(s) \rightleftharpoons trans\text{Pt}(\text{NH}_3)_2\text{I}_2(s)$$

We use the relationship $\Delta G^\circ = \Delta H^\circ - T\Delta S^\circ$ to compute ΔS°'s of the two reactions. For the *cis*-isomer

$$\Delta S^\circ = \frac{\Delta H^\circ - \Delta G^\circ}{T} = \frac{(-286.56\text{ kJ}) - (-130.25\text{ kJ})}{298.15\text{ K}} = -0.52427\text{ kJ K}^{-1}$$

For the *trans*-isomer, a similar calculation gives $\Delta S^\circ = -0.52135\text{ kJ K}^{-1}$. These ΔS°'s are the standard entropies of the products minus the standard entropies of the reactants. We represent the standard molar entropy of the *cis*-compound by x and write for the first reaction:

$$-524.27\text{ J K}^{-1} = 1(x) - 1(41.63) - 1(191.50) - 3(130.57) - 1(116.14)$$

where the numbers in parentheses are molar S° values (Appendix D) for $\text{Pt}(s)$, $\text{N}_2(g)$, $\text{H}_2(g)$, and $\text{I}_2(s)$ respectively, and the other numbers are the appropriate numbers of moles. The S° of the *cis* is $216.71\text{ J K}^{-1}\text{mol}^{-1}$. For the *trans*, a similar calculation gives $S^\circ = 219.63\text{ J K}^{-1}\text{mol}^{-1}$.

8-58 (a) The value for ΔH° computed by the student is for a process at 25°C. The value in the table is for the same change, but at a different temperature, the boiling point of the carbon tetrachloride. Interestingly, the ΔH° of vaporization of a liquid substance decreases with increasing T and becomes zero at T_c, the critical temperature.

(b) At the boiling point of the CCl_4 we have

$$\Delta S_{\text{vap}} = \frac{\Delta H_{\text{vap}}}{T} = \frac{30.0\text{ kJ mol}^{-1}}{349.65\text{ K}} = 0.0858\text{ kJ K}^{-1}\text{mol}^{-1} = 85.8\text{ J K}^{-1}\text{mol}^{-1}$$

Note that this is close to $88\text{ J K}^{-1}\text{mol}^{-1}$, the prediction of Trouton's rule.

8-60 In a reversible adiabatic expansion of a gas, the temperature decreases, so the molecules move more slowly on average. Although molecules have more positions available to them (because the volume has increased), they have fewer velocities. These effects oppose each other and lead to the number of microstates (and thus the energy) remaining constant.

Chapter 9

Chemical Equilibrium: Basic Principles and Applications to Gaseous Reactions

9-2 The point is to formulate the equilibrium expressions correctly:

(a) $\dfrac{P_{Cl_2O}^2}{P_{Cl_2}^2 P_{O_2}} = K$ (b) $\dfrac{P_{NOBr}^2}{P_{N_2} P_{O_2} P_{Br_2}} = K$ (c) $\dfrac{P_{H_2O}^4 P_{CO_2}^3}{P_{C_3H_8} P_{O_2}^5} = K$

9-4 The equation is $4\,NH_3(g) + 5\,O_2(g) \rightleftharpoons 6\,H_2O(g) + 4\,NO(g)$; the corresponding equilibrium expression is

$$\frac{P_{NO}^4 P_{H_2O}^6}{P_{NH_3}^4 P_{O_2}^5} = K$$

9-6 (a)

$$\frac{P_{COCl_2}}{P_{CO} P_{Cl_2}} = K$$

(b)

$$P_{COCl_2} = K P_{CO} P_{Cl_2} = (0.20)(0.0020)(0.00030) = 1.2 \times 10^{-7} \text{ atm}$$

9-8 An acceptable equation is $C_2H_6(g) \rightleftharpoons C_2H_2(g) + 2\,H_2(g)$. The ΔG° for this reaction is

$$\Delta G^\circ = 2\underbrace{(0)}_{H_2(g)} + 1\underbrace{(209.20)}_{C_2H_2(g)} - 1\underbrace{(-32.89)}_{C_2H_6(g)} = 242.09 \text{ kJ}$$

The $\Delta G°$ is related to the equilibrium constant of the reaction by $\Delta G° = -RT \ln K$. Hence:

$$242.09 \times 10^3 \text{ J mol}^{-1} = -RT \ln K = -(8.315 \text{ J mol}^{-1} \text{ K}^{-1})(298.15 \text{ K}) \ln K$$

$$\ln K = -97.65 \quad \text{and} \quad K = 3.9 \times 10^{-43}$$

Notice that if we choose to represent the dehydrogenation reaction with a doubled equation (coefficients 2, 2, and 4), then $\Delta G°$ comes out twice as big: the equation has doubled coefficients, and K is squared.

9-10 We compute the $\Delta G°$ of the reaction and then use the relationship $\Delta G° = -RT \ln K$.

$$\Delta G° = 2(-46.0) - 2(0) - 1(0) - 1(0) = -92.0 \text{ kJ, so}$$

$$K = 1.3 \times 10^{16} = \frac{P^2_{HNO_2}}{P_{H_2} \times P_{N_2} \times P^2_{O_2}}.$$

9-12 The equation is $F_3SSF(g) \rightleftharpoons 2\,SF_2(g)$.

$$\frac{P^2_{SF_2}}{P_{F_3SSF}} = \frac{(1.1 \times 10^{-4})^2}{(0.0484)} = 2.5 \times 10^{-7} = K$$

9-14 To calculate the equilibrium constant requires the partial pressures of all three gases at equilibrium. Set up a three-line table:

	$SbCl_5(g) \rightleftharpoons$	$SbCl_3(g) +$	$Cl_2(g)$
Initial partial pressure (atm)	x	0.0	0.0
Change in partial pressure(atm)	$-0.718x$	$+0.718x$	$+0.718x$
Equilibrium partial pressure (atm)	$0.282x$	$0.718x$	$0.718x$

What is different here is that the original partial pressure of $SbCl_5$ is not known and is represented by an x. The total pressure at equilibrium is the sum of the partial pressures of the three gases and equals 1.000 atm:

$$1.000 \text{ atm} = P_{SbCl_5} + P_{SbCl_3} + P_{Cl_2} = 0.282x + 0.718x + 0.718x$$

Solving gives $x = 0.582$ atm. The equilibrium partial pressures of the three gases are now readily computed:

$$P_{SbCl_5} = 0.164 \text{ atm}; \quad P_{SbCl_3} = P_{Cl_2} = 0.418 \text{ atm}$$

Substitution in the equilibrium expression gives

$$\frac{P_{SbCl_3} P_{Cl_2}}{P_{SbCl_5}} = \frac{(0.418)(0.418)}{(0.164)} = 1.07 = K$$

9-16 (a) The vapor density of the contents of the flask cannot change in this reaction. Before any of the NOBr has a chance to react:

$$P_{\text{NOBr}} = \left(\frac{n}{V}\right) RT = \left(2.219\frac{\text{g}}{\text{L}} \times \frac{1 \text{ mol NOBr}}{109.91 \text{ g NOBr}}\right) RT = 0.5798 \text{ atm}$$

Setting up a three-line table:

	$NOBr(g) \rightleftharpoons$	$NO(g) +$	$1/2\,Br_2(g)$
Initial partial pressure (atm)	0.5798	0.0	0.0
Change in partial pressure(atm)	$-x$	$+x$	$+1/2x$
Equilibrium partial pressure (atm)	$0.5798 - x$	x	$1/2x$

The final pressure in the flask, the sum of the three partial pressures, is 0.675 atm. It follows that

$$P_{\text{NOBr}} + P_{\text{NO}} + P_{\text{Br}_2} = (0.5798 - x) + x + 1/2x = 0.675$$

The value of x is 0.190 atm, so the three partial pressures are 0.389 atm, 0.190 atm, and 0.095 atm at equilibrium.

(b) Calculating the equilibrium constant requires only substituting the equilibrium partial pressures into the proper expression:

$$\frac{P_{\text{Br}_2}^{\frac{1}{2}} P_{\text{NO}}}{P_{\text{NOBr}}} = \frac{(0.095)^{\frac{1}{2}} (0.190)}{0.389} = 0.151 = K$$

9-18 The second equation is the first equation reversed with every coefficient divided by 6. The equilibrium constant for the second equation is accordingly the reciprocal of that of the first raised to the 1/6 power:

$$K_2 = \left(\frac{1}{32.6}\right)^{\frac{1}{6}} = 0.559$$

9-20 The equation of interest, the third equation written in the problem, is equal to the first added to the reverse of the second. Accordingly, the equilibrium constant for the third reaction equals the equilibrium constant of the first multiplied by the reciprocal of the equilibrium constant of the second:

$$K_3 = K_1(K_2)^{-1} = \frac{K_1}{K_2} = \frac{7.0 \times 10^3}{38 \times 10^3} = 0.18$$

9-22 (a) The molar mass of the isopropyl alcohol is 60.096 g mol^{-1}. The 10.00 g sample of the alcohol is 0.1664 mol and exerts a pressure of 0.6174 atm in the 10.00 L container at 452.15 K (179°C), assuming it behaves as an ideal gas. This is <u>before</u> the reaction has a chance to occur. The reaction generates acetone and hydrogen at the expense of isopropyl alcohol. Let x represent the decrease in the partial pressure of the isopropyl alcohol as the reaction comes to equilibrium. Then:

	$(CH_3)_2CHOH(g) \rightleftharpoons$	$(CH_3)_2CO(g) +$	$H_2(g)$
Init. pressure (atm)	0.6174	0	0
Change in pressure (atm)	$-x$	$+x$	$+x$
Equil. pressure (atm)	$0.6174 - x$	x	x

Substitution of the equilibrium pressures into the equilibrium constant expression gives the equation

$$\frac{x^2}{(0.6174 - x)} = K = 0.444$$

Solution of this equation gives $x = 0.3467$, so $P_{acetone} = 0.347$ atm at equilibrium.

(b) If there were no reaction the partial pressure of isopropyl alcohol would be 0.6174 atm; the actual partial pressure at equilibrium is $0.6174 - 0.3467 = 0.2707$ atm. This means that 0.562 of the isopropyl alcohol has dissociated.

9-24 The partial pressure of the hydrogen iodide is unknown at the moment that the glass vessel is filled. Let it be y. The partial pressures of the two products are both zero at the moment of filling. Attainment of equilibrium reduces the partial pressure of HI, the reactant, by some amount, say, $2x$, and increases the partial pressures of each product by x. This is summarized in the table:

	$2\,HI(g) \rightleftharpoons$	$(I_2(g) +$	$H_2(g)$
Init. pressure (atm)	y	0	0
Change in pressure (atm)	$-2x$	$+x$	$+x$
Equil. pressure (atm)	$y - 2x$	x	x

The total pressure at equilibrium, which is given in the problem, is the sum of the equilibrium partial pressures of the three components of the mixture: $(y - 2x) + x + x = 6.45$ atm. Clearly, $y = 6.45$ atm. Substitution of the equilibrium partial pressures into the expression for K gives

$$\frac{x^2}{(6.45 - 2x)^2} = K = 0.0259 \quad \text{so that} \quad \frac{x}{(6.45 - 2x)} = \sqrt{0.0259} = 0.1609$$

Solving this equation gives $x = 0.785$. Hence, $P_{H_2} = P_{I_2} = 0.79$ atm, and $P_{HI} = 4.88$ atm.

9-26 Consider the <u>reverse</u> of the reaction given in the problem and write the initial pressures, changes and final partial pressures in the usual way:

$$OF_2(g) \rightleftharpoons F_2(g) + 1/2\,O_2(g)$$

Init. pressure (atm)	y	0	0
Change in pressure (atm)	$-x$	$+x$	$+1/2x$
Equil. pressure (atm)	1.00	x	$+1/2x$

In this table, y appears as the unknown initial pressure of the $OF_2(g)$, but is not really needed because, from the equilibrium law:

$$\frac{(\frac{1}{2}x)^{\frac{1}{2}}x}{1.00} = K = \frac{1}{40.1}$$

The equilibrium constant is $1/40.1$ because the set-up is for the reverse reaction. Squaring both sides gives

$$\left(\frac{1}{40.1}\right)^2 = \left(\frac{1}{2}x\right)x^2$$

from which $x = 0.1075$. The partial pressure of the $F_2(g)$ is 0.107 atm, and the partial pressure of the $O_2(g)$ is half this, 0.054 atm.

9-28

$$2\,NO_2(g) \rightleftharpoons 2\,NO(g) + O_2$$
$$0.89 - 2x \qquad\qquad 2x \qquad\quad x$$

$$\frac{x(2x)^2}{(0.89 - 2x)^2} = 5.9 \times 10^{-13}$$

Assume that $2x$ is small compared to 0.89. Then

$$4x^3 \approx 5.9 \times 10^{-13}(0.89)^2 = 4.67 \times 10^{-13}$$

$$x \approx 4.9 \times 10^{-5}$$

We verify that $2x << 0.89$. Thus at equilibrium

$$P_{NO_2} = 0.89 \text{ atm}$$

$$P_{NO} = 2x = 9.8 \times 10^{-5} \text{ atm}$$
$$P_{O_2} = x = 4.9 \times 10^{-5} \text{ atm}$$

9-30

$$N_2(g) + 3H_2(g) \rightleftharpoons 2NH_3(g)$$
$$P_o - x \quad\quad 3P_o - 3x \quad\quad 2x$$

We have assumed inital proportions of H_2 and N_2 to be $3:1$ in the above.

$$P_{\text{total}} = P_o - x + (3P_o - 3x) + 2x = 4P_o - 2x = 1.00$$

$$P_o = 0.25 + 0.50x$$

$$P_{N_2} = P_o - x = 0.25 - 0.50x$$

$$P_{H_2} = 3P_o - 3x = 0.75 - 1.50x$$

$$P_{NH_3} = 2x$$

Inserting these into the equilibrium expression gives

$$\frac{(P_{NH_3})^2}{P_{N_2}(P_{H_2})^3} = K$$

$$\frac{(2x)^2}{(0.25 - 0.50x)(0.75 - 1.50x)^3} = 3.19 \times 10^{-4}$$

Ignoring the terms in x in the denominator and solving for x gives

$$x = 2.9 \times 10^{-3}$$

$$P_{N_2} = 0.25 - 0.50x = 0.25 \text{ atm}$$

$$P_{H_2} = 0.75 - 1.50x = 0.75 \text{ atm}$$

$$P_{NH_3} = 2x = 5.8 \times 10^{-3} \text{ atm}$$

9-32 The equilibrium constant for the reaction

$$SO_2Cl_2(g) \rightleftharpoons SO_2(g) + Cl_2(g)$$

is 2.40 at 100°C (373.15 K). Data concerning two of the three components of the equilibrium are given in terms of concentrations. Deal with this by noting that the partial pressure of any component in a gaseous mixture is related to its concentration by

$$P_A = \left(\frac{n_A}{V}\right)RT = [A]RT$$

as long as the mixture obeys Dalton's law and the ideal gas law. Substitute such concentration terms in the equilibrium expression:

$$K = \frac{P_{Cl_2}P_{SO_2}}{P_{SO_2Cl_2}} = \frac{[Cl_2]RT[SO_2]RT}{[SO_2Cl_2]RT} = \frac{[Cl_2][SO_2]}{[SO_2Cl_2]}RT$$

The partial pressures used in equilibrium expressions in the text always refer to a standard state of 1 atm. Therefore, using R in units of L atm mol^{-1}K^{-1} gives all concentrations in mol L^{-1}. The computation is:

$$2.40 = \frac{[6.9 \times 10^{-3}][SO_2]}{[3.6 \times 10^{-4}]}(0.08206)(373.15)$$

Solving gives $[SO_2] = 4.1 \times 10^{-3}$ mol L^{-1}.

9-34 (a) Although the reaction quotient Q has the form of the equilibrium expression, it equals K numerically only if true equilibrium partial pressures are inserted in it. In the situation described, the system is not at equilibrium because Q differs from K:

$$\frac{P_{SF_2}^2}{P_{F_3SSF}} = Q = \frac{(2.3 \times 10^{-4})^2}{0.0484} = 1.1 \times 10^{-6}$$

The presence of the argon, which takes no part in the reaction, is immaterial to this computation.

(b) The value of Q exceeds K, which from problem 9-12 equals 2.5×10^{-7}. The reaction will tend to proceed to the left, generating F$_3$SSF and consuming SF$_2$, until Q becomes equal to K.

9-36 According to the list of partial pressures, the reaction quotient Q at the moment of mixing the four compounds equals

$$\frac{P_{NO}P_{CO_2}}{P_{NO_2}P_{CO}} = Q = \frac{(1.4)(1.4)}{(3.4)(3.4)} = 0.17$$

The equilibrium constant for this reaction must be greater than 0.17 because the brown NO$_2$ is consumed as Q tends to become equal to K. This change increases the numerator and decreases the denominator in the equilibrium expression.

9-38 (a)

$$Q = \frac{P_{SO_2}P_{Cl_2}}{P_{SO_2Cl_2}} = \frac{0}{1.20} = 0$$

The reaction proceeds to the right.

(b)

$$SO_2Cl_2(g) \;\rightleftharpoons\; SO_2(g) \;+\; Cl_2(g)$$
$$1.20 - x \qquad\qquad x \qquad\quad x$$

$$\frac{x^2}{1.20 - x} = 2.4$$

$$x^2 + 2.4x - 2.88 = 0$$

$$x = \frac{-2.4 \pm \sqrt{(2.4)^2 + 4(2.88)}}{2} = 0.88$$

$$P_{SO_2} = P_{Cl_2} = 0.88 \text{ atm}$$

$$P_{SO_2Cl_2} = 1.20 - 0.88 = 0.32 \text{ atm}$$

(c) Net formation.

9-40 (a) If $O_2(g)$ is added to the equilibrium $SO_3(g) \rightleftharpoons SO_2(g) + 1/2\,O_2(g)$ at constant V and T, the equilibrium shifts to the left.

(b) If the mixture is compressed at constant T, the equilibrium shifts left.

(c) The equilibrium shifts left.

(d) Pumping a non-reactive gas into the equilibrium mixture at constant T and P must expand the container. Increasing the volume favors the products (right side) of this equilibrium.

(e) If an inert gas is pumped in at constant volume, the total pressure in the container rises, but the partial pressures of the reactants and products are unchanged; the equilibrium is unaffected.

9-42 (a) The reaction is endothermic.

(b) The equilibrium is shifted to the left (the reactants) when the non-reactive gas neon is admitted in such a way that the volume increases. We reason that the reactant side takes up more volume and therefore has a larger number of moles of gas than the product side. There is a net decrease in the number of gas molecules in the reaction.

9-44 Because the reaction is exothermic (giving off heat) high product yield is favored by using as low a temperature as possible. In addition, the chemical amount of gas decreases in the course of the reaction, so high total pressure favors production of methanol.

9-46 The computation of $\Delta H°$ and $\Delta S°$ for the formation of dimethyl ether from methanol follows the usual pattern:

$$\Delta H° = 1(-241.82) + 1(-184.05) - 2(-200.66) = -24.55 \text{ kJ}$$

$$\Delta S° = 1(188.72) + 1(266.27) - 2(239.70) = -24.41 \text{ J K}^{-1}$$

The equilibrium production of dimethyl ether by this means is favored by low temperatures. Pressure has little effect on K.

9-48

$$K_2 = 2900 \quad T_2 = 28°C = 301.15 \text{ K}$$
$$K_1 = 40 \quad\quad T_1 = 48°C = 321.15 \text{ K}$$

$$\ln\left(\frac{2900}{40}\right) = \frac{\Delta H°}{8.315 \text{ J mol}^{-1} \text{ K}^{-1}}\left(\frac{1}{321.15 \text{ K}} - \frac{1}{301.15 \text{ K}}\right)$$

This yields

$$\Delta H° = -172{,}000 \text{ J mol}^{-1}$$

$\Delta H°$ for the reaction as written is $-172{,}000 \text{ J} = -172 \text{ kJ}$.

At 28°C:

$$\Delta G° = -RT \ln K = -(8.315 \text{ J mol}^{-1} \text{ K}^{-1})(301.15 \text{ K})(\ln 2900) = -19{,}960 \text{ J mol}^{-1}$$

For the reaction as written, $\Delta G° = -19{,}960 \text{ J}$.

$$\Delta S° = -\frac{\Delta G° - \Delta H°}{T} = -\frac{(-19{,}960 \text{ J}) - (-172{,}200 \text{ J})}{301.15 \text{ K}} = -506 \text{ J K}^{-1}$$

9-50 The reaction is exothermic because K goes down with higher T. Use the van't Hoff equation to compute the $\Delta H°$ of the reaction. Inserting the two temperatures (on the Kelvin scale) and the two K's into the van't Hoff equation gives:

$$\ln\left(\frac{93.1}{2780}\right) = \frac{\Delta H°}{R}\left(\frac{1}{295.15} - \frac{1}{315.15}\right)$$

Solution of this equation gives $\Delta H° = -1.313 \times 10^5 \text{ J mol}^{-1}$, which means $\Delta H° = -131 \text{ kJ}$ for the reaction as written (2 mol of reactant giving 1 mol of product).

To estimate $\Delta S°$ substitute $\Delta H° = -1.313 \times 10^5 \text{ J mol}^{-1}$ into the expression:

$$\ln K = -\frac{\Delta H°}{RT} + \frac{\Delta S°}{R}$$

using either of the two K,T pairs. A common error is using units with kilojoules for $\Delta H°$ but joules for R. The correct answer is $\Delta S° = -379 \text{ J K}^{-1}$ for the reaction as written.

9-52 We apply Hess' law to the ΔH_f° data from Appendix D and calculate that $\Delta H^\circ = -98.9$ kJ for the reaction given. That is, $\Delta H^\circ = -98.9$ kJ per mole of reaction as written. From the van't Hoff equation, taking T_2 to be 823 K (550°C),

$$\ln\left(\frac{K_{823}}{K_{298}}\right) = \frac{\Delta H^\circ}{R}\left(\frac{1}{298 \text{ K}} - \frac{1}{823 \text{ K}}\right)$$

$$= \left(\frac{-98,000 \text{ J mol}^{-1}}{8.315 \text{ J mol}^{-1} \text{ K}^{-1}}\right)\left(\frac{1}{298 \text{ K}} - \frac{1}{823 \text{ K}}\right) = -25.46$$

$$\frac{K_{823}}{K_{298}} = e^{-25.46} = 8.8 \times 10^{-12}$$

$$K_{823} = (8.8 \times 10^{-12}) \times (2.6 \times 10^{12}) = 23$$

This result differs somewhat from the experimental K_{823} because the temperature dependence of ΔH° and ΔS° was neglected over quite a large range of temperature. Because the reaction is exothermic, an increase in temperature reduces the equilibrium constant.

9-54 (a) The problem is easily solved by substitution in the Clausius-Clapeyron equation:

$$\ln\left(\frac{P_2}{P_1}\right) = \frac{-\Delta H_{vap}}{R}\left(\frac{1}{T_2} - \frac{1}{T_1}\right)$$

The problem gives P_1 and T_1 as 0.1316 atm and 343.25 K and P_2 and T_2 as 0.5263 atm and 373.95 K. Solution for ΔH_{vap} gives 48.19 kJ mol^{-1}.

(b) We now set P_2 equal to 1.000 atm (at the normal boiling point) and T_2 equal to T_b:

$$\ln\left(\frac{1.000}{0.1316}\right) = \frac{-48.19 \times 10^3 \text{ kJ mol}^{-1}}{8.315 \text{ J mol}^{-1} \text{ K}^{-1}}\left(\frac{1}{T_b} - \frac{1}{343.25 \text{ K}}\right)$$

Solving for T_b gives $T_b = 390.1$ K, which is equivalent to 117.0°C. Note that substitution of the other P, T pair (0.5263 atm and 373.95 K) for P_2 and T_2 in the equation gives the same answer.

9-56 The synthetic reaction is:

$$CO(g) + 2\,H_2(g) \rightleftharpoons CH_3OH(g) \quad \text{with} \quad \frac{P_{CH_3OH}}{P_{CO}P_{H_2}^2} = 6.08 \times 10^{-3} \text{ at } 225°C$$

Let the equilibrium partial pressure of the H_2 be $2x$; then $P_{CO} = x$. Meanwhile, the partial pressure of the methanol equals 0.500 atm. Substitution gives:

$$\frac{0.500}{x(2x)^2} = 6.08 \times 10^{-3}$$

Solving for x gives 2.74. The equilibrium partial pressure of the CO is 2.74 atm, and that of the H_2 is 5.48 atm.

9-58 (a) To calculate the degree of conversion of the *t*-butanol requires knowledge of its partial pressure at equilibrium. Set up a three-line table:

	$(CH_3)_3COH(g) \rightleftharpoons$	$(CH_3)_2CCH_2(g)\ +$	$H_2O(g)$
Init. pressure (atm)	0.100	0.0	0.0
Change in pressure (atm)	$-x$	$+x$	$+x$
Equil. pressure (atm)	$0.100 - x$	x	x

Putting the equilibrium values into the equilibrium law gives

$$\frac{x^2}{(0.100 - x)} = K = 2.42$$

from which x is 0.0962 atm. The fraction of *t*-butanol that is converted at equilibrium is 0.962.

(b) A similar calculation gives $x = 2.473$, so the fraction converted is 0.495.

9-60 (a) Let y stand for the equilibrium partial pressure of the acetic acid dimer and x stand for the equilibrium partial pressure of the monomer. According to the problem, the sum of these two partial pressures is 0.725 atm, that is, $x + y = 0.725$. Also, the two partial pressures are related by the equilibrium law

$$\frac{P_{dimer}}{P^2_{monomer}} = \frac{y}{x^2} = 3.72$$

There are thus two relationships governing two unknowns. Eliminating x and solving (by means of the quadratic formula) gives $y = 0.398$ atm. This is the equilibrium partial pressure of the dimer.

(b) Solving for x in the previous part gives the equilibrium partial pressure of the monomeric acetic acid as 0.327 atm. If none of the acetic acid were dimerized, the partial pressure of the monomer would be $0.327 + (2 \times 0.398) = 1.123$ atm. The pressure of monomer that is actually present is 0.327 atm. This is 29.1% of 1.123 atm; it follows that 70.9% of the acetic acid is present as the dimer.

9-62

$$\frac{0.800 \text{ g}}{80.063 \text{ g mol}^{-1}} = 9.99 \times 10^{-3} \text{ mol } SO_3$$

Its initial partial pressure is

$$P = \frac{nRT}{V} = \frac{(9.99 \times 10^{-3} \text{ mol})(0.08206 \text{ L atm mol}^{-1}\text{K}^{-1})(900 \text{ K})}{0.1000 \text{ L}} = 7.38 \text{ atm}$$

$$\tfrac{1}{2}O_2(g) \;+\; SO_2(g) \;\rightleftharpoons\; SO_3(g)$$
$$\quad\tfrac{1}{2}x \qquad\qquad x \qquad\qquad 7.38 - x$$

$$\frac{7.38 - x}{x\left(\tfrac{1}{2}x\right)^{1/2}} = K = 0.587$$

To simplify this, mulitply both sides by $(\tfrac{1}{2})^{1/2}$ and then take the inverse of both sides. This leaves

$$\frac{x^{3/2}}{7.38 - x} = 2.41$$

If we set x to zero in the denominator, we find

$$x^{3/2} \approx (2.41)(7.38) = 17.8$$

$$x \approx (17.8)^{2/3} = 6.81$$

Successive iterations give 1.23, 6.03, ...

This is clearly converging very slowly. It is easier to guess values x between 1.23 and 6.03, calculate the left side, and compare it to the right side. This gives

Guess	Left side
4	2.37
4.1	2.53
4.02	2.399
4.03	2.415

Thus $x = 4.03$.

$$P_{O_2} = \frac{1}{2}x = 2.01 \text{ atm}$$

9-64 A process can be quite exothermic, but still non-spontaneous. At constant T and P the condition for spontaneity is the sign of ΔG. Spontaneity is thus a compromise between two terms, an enthalpy term ΔH and an entropy term $-T\Delta S$. If ΔS is negative, then $-T\Delta S$ is positive. A large enough T forces ΔG to become positive. The more negative the ΔS, then the lower the required temperature. The fact that helium is only slightly soluble in $H_2O(l)$ means that $-T\Delta S$ for $He(g) \rightarrow He(aq)$ overmatches the negative ΔH. That is only possible if ΔS is negative.

9-66 The initial chemical amount of methanol is 0.1473 mol, so its initial partial pressure, calculated from the ideal gas law, is 6.324 atm. Suppose this is reduced by x atm at equilibrium, giving an equilibrium partial pressure of $2x$ atm for $H_2(g)$. According to Graham's law of effusion (Section 3-6) the ratio of effusion rates of two gases from a single container is

$$\frac{\text{rate of effusion of } H_2}{\text{rate of effusion of } CH_3OH} = \frac{N_{H_2}}{N_{CH_2OH}} \sqrt{\frac{\mathcal{M}_{CH_3OH}}{\mathcal{M}_{H_2}}}$$

The problem states that the ratio on the left side of this equation is 33, and the molar masses are easily calculated. The ratio of the number of molecules is equal to the ratio of partial pressures, because volume and temperature are the same. Thus,

$$33 = \frac{2x}{6.324 - x} \sqrt{\frac{32.04}{2.016}}$$

Solving this gives $x = 5.093$. The equilibrium partial pressures are thus $P_{CO} = x = 5.093$ atm, $P_{H_2} = 2x = 10.19$ atm, and $P_{CH_3OH} = 6.324 - x = 1.231$ atm. The equilibrium constant is

$$K = \frac{P_{CO} \left(P_{H_2}\right)^2}{P_{CH_3OH}} = 430$$

Chapter 10

Acid-Base Equilibria

10-2 All five can act as Brønsted-Lowry bases. Their conjugate acids are HF, HSO_4^-, OH^-, H_2O and H_3O^+.

10-4 $NH_3(aq) + CH_3COOH(aq) \rightarrow NH_4^+(aq) + CH_3COO^-(aq)$

10-6 (a) The slag-forming reaction is $CaO + SiO_2 \rightarrow CaSiO_3$.

(b) The CaO is a Lewis base and the SiO_2 is a Lewis acid.

10-8 (a) This problem follows up the extension of the acid-base concepts suggested by problem 10-6. In the Brønsted system, the acid is the *donor* of a *positively* charged particle (the hydrogen ion); in this system (called the Lux-Flood acid-base system) the acid is the *acceptor* of a *negatively* charged particle (the oxide ion), and the base is the donor of the oxide ion.

(b) In the first reaction CaO donates O^{2-} to SiO_2 so CaO is the base, and SiO_2 is the acid. In the second reaction the SiO_2 again accepts an O^{2-} and is again the acid. It accepts the O^{2-} from Ca_2SiO_4, which is the base. In the third reaction, the CaO donates O^{2-} to Ca_2SiO_4. The latter serves as an acid in this reaction, the opposite of its role in the second reaction.

10-10 (a) The anhydride is As_2O_5, diarsenic pentaoxide.
(b) The anhydride of this compound, which is called molybdic acid, is molybdenum(VI) oxide, MoO_3.
(c) The anhydride is Rb_2O, rubidium oxide.
(d) The anhydride is SO_2, sulfur dioxide.

10-12
$$ZnO(s) + 2\,HCl(aq) \rightarrow Zn^{2+}(aq) + 2\,Cl^-(aq) + H_2O(l)$$
$$ZnO(s) + 2\,NaOH(aq) + H_2O(l) \rightarrow Zn(OH)_4^{2-}(aq) + 2\,Na^+(aq)$$

10-14 Calculate the concentration of H_3O^+ in the bleach using the relationship $K_w = [OH^-][H_3O^+]$ with $K_w = 1.0 \times 10^{-14}$. The pH is the negative logarithm of the $[H_3O^+]$; the answer is pH $= 12.56$.

10-16 For H_3O^+ ion, the range is 3.5×10^{-8} to 4.5×10^{-8} M. For OH^- ion the range is 2.2×10^{-7} to 2.8×10^{-7} M.

10-18 The concentration of hydronium ion is computed from the definition of pH

$$[H_3O^+] = 10^{-7.4} = 4 \times 10^{-8} \text{ M}$$

To determine the concentration of OH^- we use the equilibrium expression for the autoionization of water

$$[OH^-] = \frac{K_w}{[H_3O^+]} = \frac{2.4 \times 10^{-14}}{4 \times 10^{-8}} = 6 \times 10^{-7} \text{ M}$$

10-20 Raising the pH increases the concentration of OH^- ions. This makes the direct substitution reaction much more feasible

$$(CH_3)_3CCl + OH^- \rightarrow (CH_3)_3COH + Cl^-$$

10-22 (a) The ionization of niacin proceeds by the reaction

$$C_5H_4NCOOH(aq) + H_2O(l) \rightleftharpoons C_5H_4NCOO^-(aq) + H_3O^+(aq)$$

(b) In aqueous solutions, the product of K_a of an acid and K_b of its conjugate base is K_w. Hence

$$K_b = \frac{K_w}{K_a} = \frac{1.0 \times 10^{-14}}{1.5 \times 10^{-5}} = 6.7 \times 10^{-10}$$

(c) The K_a for the pyridinium ion appears in Table 10.2. It is 5.6×10^{-6}, smaller than the K_a of niacin. Niacin is a stronger acid than pyridinium ion, which means that its conjugate base is a weaker base than pyridine.

10-24 The equation given in the problem is the sum of the chemical equation for the third acid ionization reaction of phosphoric acid

$$HPO_4^{2-}(aq) + H_2O(l) \rightleftharpoons H_3O^+(aq) + PO_4^{3-}(aq)$$

and the reverse of the equation for the first acid ionization of carbonic acid (H_2CO_3)

$$H_2CO_3(aq) + H_2O(l) \rightleftharpoons H_3O^+(aq) + HCO_3^-(aq)$$

Hence, the desired equilibrium constant is

$$K = \frac{K_{a3,H_3PO_4}}{K_{a1,H_2CO_3}} = \frac{2.2 \times 10^{-13}}{4.3 \times 10^{-7}} = 5.1 \times 10^{-7}$$

H_2CO_3 is the stronger acid, and PO_4^{3-} is the stronger base.

10-26 (a) According to Figure 10.8, cresol red starts its change from its basic to acidic color at a pH of 8.8, but thymolphthalein starts at the less acidic pH of 10.6. The color changes mark the pH at which the concentrations of the acid forms of the indicators start to become important. The basic form of the thymolphthalein requires a lesser concentration of H_3O^+ to cause it to change to the acid form. It is accordingly a stronger base than the basic form of cresol red.

(b) Clearly, 8.8 < pH < 9.4 (see Figure 10.8.)

10-28 The molar mass $\mathcal{M}$ of ascorbic acid is 176.126 g mol^{-1}. The concentration of ascorbic acid in the 100 mL of water is its chemical amount, 2.839×10^{-3} mol (computed by dividing 500×10^{-3} g by $\mathcal{M}$) itself divided by the volume of the solution (0.100 L). It is 0.0284 M.

The acid-ionization equilibrium is

$$HC_6H_7O_6(aq) + H_2O(l) \rightleftharpoons H_3O^+(aq) + C_6H_7O_6^-(aq)$$

for which the equilibrium expression is

$$\frac{[H_3O^+][C_6H_7O_6^-]}{[HC_6H_7O_6]} = K_a = 8.0 \times 10^{-5}$$

$$\frac{x^2}{0.0284 - x} = 8.0 \times 10^{-5}$$

Solving the equation gives $x = 1.47 \times 10^{-3}$. The pH of the solution is $-\log(1.47 \times 10^{-3}) = 2.83$.

10-30 (a) The K_a of propionic acid is 1.34×10^{-5} (see Table 10.2 of the text). Following the pattern of Example 10.3 gives the concentration of H_3O^+ as 2.16×10^{-3} M, for a pH of 2.664.

(b) A solution of formic acid of pH 2.664 is to be prepared. The K_a of formic acid is 1.77×10^{-4}. Because formic acid is a stronger acid than propionic acid, less of it than propionic acid is required to lower the pH of a liter of pure water from 7.0 to 2.664, a pH corresponding to $[H_3O^+] = 2.16 \times 10^{-3}$ M. Let the required concentration of formic acid be c. Then, at equilibrium

$$1.77 \times 10^{-4} = \frac{(2.16 \times 10^{-3})(2.16 \times 10^{-3})}{(c - 2.16 \times 10^{-3})}$$

Solving gives $c = 2.85 \times 10^{-2}$; the required concentration of formic acid is only 2.85×10^{-2}M.

10-32 If the equilibrium concentration of H_3O^+ is x, then

$$[C_6F_5COO^-] = x \quad \text{and} \quad [C_6F_5COOH] = 0.100 - x$$

Substitution into the equilibrium expression gives

$$\frac{[H_3O^+][C_6F_5COO^-]}{[C_6F_5COOH]} = K_a = 0.033 = \frac{x^2}{(0.100 - x)}$$

This equation can be solved by the quadratic formula or by iteration to give $x = 0.0433$. The pH is $-\log_{10}(0.0433) = 1.36$.

10-34 From the given pH of the solution (2.42), we have $[H_3O^+] = 3.8 \times 10^{-3}$ M. This is also essentially the concentration of the 2-germaacetate ion at equilibrium; the concentration of the un-ionized 2-germaacetic acid is then $0.050 - (3.8 \times 10^{-3}) = 0.0462$ M. Substitution of these values into the equilibrium expression gives $K_a = 3.1 \times 10^{-4}$.

10-36 Write methylamine as MA. Its initial concentration is $(0.070 \text{ mol}/0.8000 \text{ L}) = 0.0875$ M. It reacts according to a standard base dissociation:

$$MA(aq) + H_2O(l) \rightleftharpoons HMA^+(aq) + OH^-(aq)$$

If x mol L^{-1} of MA dissociates, giving $[HMA^+] = [OH^-] = x$, then the equilibrium expression becomes

$$\frac{x^2}{0.0875 - x} = K_b = 4.4 \times 10^{-4}$$

$$x = 6.0 \times 10^{-3} = [OH^-]$$

$$[H_3O^+] = \frac{K_w}{[OH^-]} = 1.7 \times 10^{-12} \text{ M}$$

$$pH = 11.78$$

10-38

$$CN^-(aq) + H_2O(l) \rightleftharpoons HCN(aq) + OH^-(aq)$$

$$\frac{[OH^-][HCN]}{[CN^-]} = K_b = 2.0 \times 10^{-5}$$

$$pOH = 14.00 - 11.50 = 2.50; \ [OH^-] = [HCN] = 10^{-2.50} = 3.20 \times 10^{-3}$$

$$[CN^-] = \frac{[OH^-][HCN]}{K_b} = 0.49 \ M$$

10-40 The solution contains 75.00×0.0460 mmol of a strong acid ($HClO_4$) and is treated with 150.00×0.0230 mmol of a strong base (KOH). The chemical amounts of the acid and base are equal so the two exactly neutralize to water containing K^+ and ClO_4^- ions. Neither of the ions acts detectably as an acid or base in water so the pH of the solution equals 7.0.

10-42 $HBr < NH_4I < NaCl < KF < LiOH$

10-44 The reaction of the weak base "bis" in water can be represented:

$$bis(aq) + H_2O(l) \rightleftharpoons bisH^+(aq) + OH^-(aq)$$

for which the equilibrium expression is

$$K_b = \frac{[OH^-][bisH^+]}{[bis]} = 10^{-8.8}$$

The concentrations of the bis and its conjugate acid are essentially equal under the conditions described in the problem because the amount of the HCl that has been added is just enough to convert half of the bis to $bisH^+$ and leave half unreacted. It is true that both of these species then react with water, but the changes in amount caused by these interactions are negligible. We therefore have $K_b = [OH^-]$ and $pK_b = pOH$. The pOH is 8.8, and the pH is $14.0 - 8.8 = 5.2$.

10-46 The acid ionization of sulfanilic acid has the equilibrium law

$$\frac{[NH_2C_6H_4SO_3^-][H_3O^+]}{[NH_2C_6H_4SO_3H]} = K_a = 5.9 \times 10^{-4}$$

(a) The pH is computed by substituting 0.20 M - x for the concentration of sulfanilic acid and 0.13 M + x for the concentration of sulfanilate ion, solving for $x = [H_3O^+]$ and taking the negative logarithm. The pH is 3.04.

(b) Adding the HCl converts 0.040 mol of sulfanilate ion to its conjugate acid, sulfanilic acid. The concentrations of the two are 0.24 M - x and 0.09 M + x, respectively. The pH is 2.80.

10-48 The pK_a for the first acid ionization of carbonic acid is 6.37. The pK_a for the second acid ionization of phosphoric acid is 7.21. From the standpoint of suitability for pH control, the $H_2PO_4^-/HPO_4^{2-}$ system would be better because the applicable pK_a is closer to the desired pH.

10-50

$$[H_3O^+] = 10^{-9.60} = 2.5 \times 10^{-10}$$

$$\frac{[H_3O^-][CN^-]}{[HCN]} = K_a = 6.17 \times 10^{-10}$$

$$\frac{[CN^-]}{[HCN]} = \frac{K_a}{H_3O^+} = 2.46$$

Initially, we have $(0.400 \text{ L})(0.0800 \text{ mol L}^{-1} = 0.0320 \text{ mol CN}^-$. In the final solution, the ratio of the concentrations equals the ratio of the number of moles. If $x =$ moles HCN, then $0.320 - x =$ moles CN^-.

$$\frac{0.0320 - x}{x} = 2.46$$

$$x = 0.00926 \text{ mol}$$

This is the chemical amount of H_3O^+ that must be added, because each mole of H_3O^+ gives one of HCN. Then

$$\text{Volume HCl} = \frac{0.00926 \text{ mol}}{0.100 \text{ mol L}^{-1}} = 0.0926 \text{ L} = 92.6 \text{ mL}$$

10-52 This is the titration of a strong acid with a strong base. Before any base is added, the concentration of H_3O^+ in the solution is 0.1439 M, because the HBr is completely ionized, and the pH is 0.842.

It is easy to verify that it requires 31.14 mL of 0.1219 M NaOH to titrate the 26.38 mL of 0.1439 M HBr to the equivalence point. At the equivalence point, the pH is 7.00. When the titration is 1.00 mL short of equivalence 30.14 mL of NaOH has been added. The unreacted H_3O^+ comprises 0.122 mmol (the difference between the chemical amount of H_3O^+ originally present and the chemical amount of NaOH added), and the volume of the solution is 26.38 mL + 30.14 mL = 56.52 mL. The concentration of the H_3O^+ is its chemical amount divided by this volume or 2.16×10^{-3} M, and the pH is 2.666.

When 32.14 mL of NaOH has been added, the concentration of OH^- is the chemical amount of unreacted OH^- divided by the volume of the solution. This is 0.122 mmol divided by 58.52 mL or 2.085×10^{-3} mol L^{-1}. The pOH is 2.681, and the pH is $14.000 - $ pOH or 11.319.

10-54 (a)

$$\frac{x^2}{0.1000 - x} = 1.4 \times 10^{-3}$$

$$x = 1.12 \times 10^{-2} \text{ M} = [\text{H}_3\text{O}^+] \qquad \text{pH} = 1.95$$

(b)

$$\frac{\left(\frac{5.00 \times 0.1000 \text{ mmol}}{55.00 \text{ mL}} + x\right) x}{\left(\frac{4.50 \text{ mmol}}{55.00 \text{ mL}} - x\right)} = 1.4 \times 10^{-3}$$

$$\frac{(9.091 \times 10^{-3} + x)x}{8.182 \times 10^{-2} - x} = 1.4 \times 10^{-3}$$

$$x^2 + 1.049 \times 10^{-2}x - 1.145 \times 10^{-4} = 0$$

$$x = 6.67 \times 10^{-3} = [\text{H}_3\text{O}^+] \qquad \text{pH} = 2.18$$

(c)

$$\frac{\left(\frac{2.50 \text{ mmol}}{75.00 \text{ mL}} + x\right) x}{\left(\frac{2.50 \text{ mmol}}{75.00 \text{ mL}} - x\right)} = 1.4 \times 10^{-3}$$

$$\frac{(0.0333 + x)x}{0.0333 - x} = 1.4 \times 10^{-3}$$

$$x = 1.3 \times 10^{-3} \qquad \text{pH} = 2.89$$

(d)

$$\frac{\left(\frac{4.90 \text{ mmol}}{99.00 \text{ mL}} + x\right) x}{\left(\frac{0.10 \text{ mmol}}{99.00 \text{ mL}} - x\right)} = 1.4 \times 10^{-3}$$

$$x = 2.8 \times 10^{-5} = [\text{H}_3\text{O}^+] \qquad \text{pH} = 4.56$$

(e)

$$\frac{\left(\frac{4.99 \text{ mmol}}{99.90 \text{ mL}} + x\right) x}{\left(\frac{0.010 \text{ mmol}}{99.90 \text{ mL}} - x\right)} = 1.4 \times 10^{-3}$$

$$x = 2.7 \times 10^{-6} = [\text{H}_3\text{O}^+] \qquad \text{pH} = 5.56$$

(f) At the equivalence point, we have a 0.0500 M $\text{CH}_2\text{ClCOO}^-$ solution.

$$\text{CH}_2\text{ClOO}^-(aq) + \text{H}_2\text{O}(l) \rightleftharpoons \text{CH}_2\text{ClCOOH}(aq) + \text{OH}^-(aq)$$

$$[\text{CH}_2\text{ClCOOH}] = [\text{OH}^-] = x$$

$$\frac{x^2}{0.0500 - x} = K_b = \frac{1.0 \times 10^{-14}}{1.4 \times 10^{-3}} = 7.14 \times 10^{-12}$$

$$x = 6.0 \times 10^{-7} = [OH^-]$$

$$pOH = 6.22 \qquad pH = 7.78$$

(g) 0.10 mL NaOH beyond the equivalence point.

$$\frac{(0.10 \text{ mL})(0.1000 \text{ M})}{100.10 \text{ mL}} = 1.0 \times 10^{-4} \text{ M} = [OH^-]$$

$$pOH = 4.00 \qquad pH = 10.00$$

(h)

$$\frac{(5.00 \text{ mL})(0.1000 \text{ M})}{105.00 \text{ mL}} = 4.76 \times 10^{-3} \text{ M} = [OH^-]$$

$$pOH = 2.32 \qquad pH = 11.68$$

10-56 Before any titrating acid is added, we have a 0.175 M solution of aqueous NH_3. Ammonia is a base:

$$NH_3(aq) + H_2O(l) \rightleftharpoons NH_4^+(aq) + OH^-(aq)$$

and we have the equilibrium law

$$\frac{[NH_4^+][OH^-]}{[NH_3]} = K_b = 1.8 \times 10^{-5}$$

If the equilibrium concentration of NH_4^+ is x, then the equilibrium concentration of NH_3 is $0.175 - x$ and that of OH^- is x. The other sources of OH^- are negligible. Substitution and solving for x (by iteration) gives $[OH^-] = 1.766 \times 10^{-3}$ M and a pH of 11.25.

The text (page 343) shows that in the titration of a weak acid with a strong base, the pH at the half-equivalence point is very close to the pK_a of the weak acid: $pK_a = pH$. We construct the analogous relationship $pK_b = pOH$ to apply to the titration of a weak base with a strong acid. Then, $pH = 14.00 - pOH = 14.00 - 4.745 = 9.26$

It requires $(0.175/0.106)(140.0) = 231.13$ mL to titrate the solution to the equivalence point, at which point the total volume of the solution is 371.13 mL. The solution is in effect dilute aqueous NH_4Cl in which the concentration of the NH_4^+ ion is nominally $0.175(140.00/371.13) = 0.0660$ M. This concentration is lowered slightly by the reaction of the NH_4^+ ion with water to generate H_3O^+ and NH_3. The K_a for this interaction is

$$\frac{K_w}{K_b} = \frac{1.0 \times 10^{-14}}{1.8 \times 10^{-5}} = 5.56 \times 10^{-10}$$

Substituting in the usual way into the equilibrium expression for this acid ionization gives $[H_3O^+] = 6.06 \times 10^{-6}$ M, pH 5.22.

When the titration is 1.00 mL past the equivalence point, the NH_4^+ is still present but is completely overshadowed as a source of H_3O^+ by the excess $HCl(aq)$ and we need consider only the HCl. The first 231.13 mL of 0.106 M HCl were neutralized, so the effective concentration of HCl in the solution is the amount of HCl contributed by the last 1.00 mL divided by the total volume of the solution. This is $(1.00/372.13)(0.106) = 2.85 \times 10^{-4}$ M. This is also the concentration of H_3O^+ so the pH is 3.55.

10-58 The total chemical amount of HCl added is 18.393 mmol. The total chemical amount of NaOH added is 7.917 mmol. These numbers, which are the concentrations of the acid and base multiplied by the respective total volumes in mL, are clearly not equal. The tablet supplies the extra base needed to neutralize $18.393 - 7.917 = 10.476$ mmol of acid. The reaction is

$$CaCO_3(s) + 2\,H_3O^+(aq) \rightleftharpoons Ca^{2+}(aq) + CO_2(g) + 3H_2O(l)$$

Each mole of $CaCO_3$ takes up 2 mol of H_3O^+ so there is in the tablet $((10.476/2) - 5.238)$ mmol of $CaCO_3$. The molar mass $\mathcal{M}$ of $CaCO_3$ is 100.09 g mol^{-1} so the tablet contains 0.5243 g of $CaCO_3$. This is 39.54% of the total.

10-60 At the end-point of her titration, the chemist will have a solution of $NH_4^+(aq)$ ion that is about 0.050 M. This Brønsted-Lowry acid will make the solution acidic by the reaction

$$NH_4^+(aq) + H_2O(l) \rightleftharpoons NH_3(aq) + H_3O^+(aq)$$

The K_a for this equilibrium is 5.6×10^{-10}. If we let x equal the equilibrium concentration of the H_3O^+ and NH_3, then

$$5.6 \times 10^{-10} = \frac{x^2}{0.050 - x}$$

The x in this equation is about 5.3×10^{-6} so the pH is about 5.3. An appropriate indicator would be methyl red.

10-62 The initial concentration of cacodylate ion is 0.1000 M. The 50.00 mL aliquot of this solution contains 0.005000 mol of this ion. The 29.55 mL of 0.100 M HCl contains 0.002955 mol of HCl. When the acid is added it is completely converted to 0.002955 mol of cacodylic acid. Some cacodylate ion, $0.005000 - 0.002955 = 0.002045$ mol, is left in excess. The equilibrium expression for the acid ionization of cacodylic acid is:

$$\frac{[\text{cacodylate}][H_3O^+]}{[\text{cacodylic acid}]} = K_a$$

All the quantities on the left are known:

$$[\text{cacodylate}] = \frac{0.002045}{0.07955} \text{ M} \qquad [\text{cacodylic acid}] = \frac{0.002955}{0.07955} \text{ M}$$

$$[H_3O^+] = 1.0 \times 10^{-6} \text{ M}$$

Substitution gives $K_a = 6.9 \times 10^{-7}$.

10-64 The successive equilibrium constants for the two-step ionization of phthalic acid differ by a factor of several thousand. Because the amounts of HPh^- and H_3O^+ produced by the first step are far larger than the amount consumed (in the case of HPh^-) or produced (in the case of H_3O^+) by the second step, the steps may be considered separately. Also, the autoionization of water is a negligible source of H_3O^+ in this solution. The first stage in the ionization provides the expression

$$\frac{[H_3O^+][HPh^-]}{[H_2Ph]} = K_{a1} = 1.26 \times 10^{-3} = \frac{x^2}{0.0100 - x}$$

where x is the equilibrium concentration of H_3O^+. Solving the resulting equation gives $x = 2.98 \times 10^{-3}$. The equilibrium concentrations of HPh^- and H_3O^+ are 2.98×10^{-3}M. That of H_2Ph is $0.0100 - x = 0.0070$ M.

The second stage of ionization is governed by the expression

$$\frac{[H_3O^+][Ph^{2-}]}{[HPh^-]} = K_{a2} = 3.10 \times 10^{-6} = \frac{y(2.98 \times 10^{-3})}{2.98 \times 10^{-3}}$$

where y is the equilibrium concentration of the Ph^{2-} ion. $[Ph^{2-}] = 3.10 \times 10^{-6}$M.

10-66 This problem resembles problem 10-64 in that it treats a two-stage process in which the successive equilibrium constants differ so greatly that the stages can be treated separately. Now however, a weak base, the oxalate ion, is in solution. It reacts according to the equations

$$C_2O_4^{2-}(aq) + H_2O(l) \rightleftharpoons HC_2O_4^-(aq) + OH^-(aq) \qquad K_{b1} = K_w/K_{a2}$$
$$HC_2O_4^-(aq) + H_2O(l) \rightleftharpoons H_2C_2O_4(aq) + OH^-(aq) \qquad K_{b2} = K_w/K_{a1}$$

Let the final concentration of $HC_2O_4^-$ equal y. This is also the final concentration of OH^- because the first reaction is the only important source of OH^-. The concentration of $C_2O_4^{2-}$ at equilibrium is then $0.10 - y$. Therefore

$$\frac{y^2}{0.10 - y} = \frac{1.0 \times 10^{-14}}{6.4 \times 10^{-5}}$$

Thus, $y = 4.0 \times 10^{-6}$, and $[OH^-] = [HC_2O_4^-] = 4.0 \times 10^{-6}$ M. The hydrogen oxalate ion $(HC_2O_4^-)$ itself can serve as a weak base, as represented in the second chemical equation above. For that equation we have

$$\frac{[OH^-][H_2C_2O_4]}{[HC_2O_4^-]} = K_{b2} = \frac{K_w}{K_{a1}} = \frac{1.0 \times 10^{-14}}{5.9 \times 10^{-2}}$$

Insertion of the known equilibrium concentrations of OH^- and $HC_2O_4^-$ gives $[H_2C_2O_4] = 1.7 \times 10^{-13}$ M. At the moment of mixing, the concentration of the oxalate ion was 0.10 M. Some of this oxalate reacted, but not much—there is a tiny concentration of hydrogen oxalate ion (only 4.0×10^{-6} M) and an incredibly tiny concentration of oxalic acid (1.7×10^{-13} M). The final concentration of oxalate ion is still 0.10 M. It is clear how very slight the second reaction is as a source of OH^-.

10-68 Dissolved carbonates are in the rain drop as either carbonic acid, hydrogen carbonate ion or carbonate ion:

$$3.6 \times 10^{-5} = [H_2CO_3] + [HCO_3^-] + [CO_3^{2-}]$$

where all of the quantities on the right are to be determined. The existence of the acid ionization equilibria of carbonic acid provides additional relationships among these quantities:

$$\frac{[HCO_3^-][H_3O^+]}{[H_2CO_3]} = K_{a1} = 4.3 \times 10^{-7}$$

$$\frac{[CO_3^{2-}][H_3O^+]}{[HCO_3^-]} = K_{a2} = 4.8 \times 10^{-11}$$

The $[H_3O^+]$ is 1.0×10^{-4} M in these equations, so we have three equations in three unknowns. The easiest way to solve this system of equations is to observe that $[CO_3^{2-}]$ is likely to be quite small—almost all of the dissolved carbonate is tied up with hydrogen ion because hydrogen ion is abundant in this acidic solution. Setting $[CO_3^{2-}]$ to zero gives:

$$3.6 \times 10^{-5} \approx [H_2CO_3] + [HCO_3^-]$$

$$\frac{[HCO_3^-](1.0 \times 10^{-4})}{[H_2CO_3]} \approx 4.3 \times 10^{-7}$$

Solving this pair of equations gives $[H_2CO_3] = 3.6 \times 10^{-5}$ M and $[HCO_3^-] = 1.55 \times 10^{-7}$ M. Now, substitution in the K_{a2} expression gives $[CO_3^{2-}] = 7.4 \times 10^{-14}$ M. Almost all of the carbonate is present as carbonic acid, a small amount is present as hydrogen carbonate ion, and a vanishingly tiny fraction is present as CO_3^{2-} ion.

10-70

$$\frac{40.0 \text{ g L}^{-1}}{60.05 \text{ g mol}^{-1}} = 0.666 \text{ M}$$

After dilution, $c_a = [CH_3COOH]_0 = 6.66 \times 10^{-7}$ M

$$c_b = 0$$

The cubic equation in Section 10.8 gives, with $x = [H_3O^+]$,

$$x^3 + K_a x^2 - (K_w + c_a K_a)x - K_a K_w = 0$$

$$x^3 + 1.76 \times 10^{-5} x^2 - 1.173 \times 10^{-11} x - 1.76 \times 10^{-19} = 0$$

This can be solved by choosing a series of values for x and finding where the function passes through zero (see Appendix C). Values of $x > 10^{-7}$ are the relevant ones.

x	$f(x)$
10^{-7}	-1.17×10^{-18}
10^{-6}	6.69×10^{-18}
3×10^{-7}	-2.08×10^{-18}
7×10^{-7}	5.78×10^{-19}
6×10^{-7}	-6.63×10^{-19}
6.5×10^{-7}	-9×10^{-20}
6.57×10^{-7}	-3.8×10^{-21}

$$[H_3O^+] = 6.57 \times 10^{-7} \text{ M} \quad pH = 6.18$$

10-72 (a)

$$C_{20}H_{24}O_2N_2 + H_2O \rightleftharpoons C_{20}H_{24}O_2N_2H^+ + OH^-$$
$$\text{(Q)} \qquad\qquad\qquad \text{(HQ}^+\text{)}$$

$$K_{b1} = 3.31 \times 10^{-6}$$

Initially, $\dfrac{1.622 \text{ g}}{(324.4 \text{ g mol}^{-1})(0.100 \text{ L})} = 0.0500 \text{ M}$

$$x = [OH^-] = [HQ^+]$$

$$\frac{x^2}{0.0500 - x} = 3.31 \times 10^{-6}$$

$$x = 4.05 \times 10^{-4} \text{ M}$$

$$[H_3O^+] = \frac{1.00 \times 10^{-14}}{4.05 \times 10^{-4}} = 2.47 \times 10^{-11} \text{ M} \qquad \text{pH} = 10.61$$

(b) At 25.00 mL HCl added, $[Q]_0 = [HQ^+]_0$

$$\frac{[HQ^+][OH^-]}{[Q]} = \frac{(0.0200 + x)x}{0.0200 - x} = 3.31 \times 10^{-6}$$

$$x = [OH^-] = 3.31 \times 10^{-6} \qquad \text{pH} = 8.52$$

(c) This is an amphoteric equilibrium problem, equivalent to preparing a 0.0333 M solution of HQ^+.

$$K_{a1} = K_w/K_{b2} = 7.41 \times 10^{-5} << [HQ^+] = 0.033$$

$$K_{a2} = K_w/K_{b1} = 3.02 \times 10^{-9}; \quad K_{a2}[HQ^+] = 1.0 \times 10^{-10} >> K_w$$

From the equation on p. 351,

$$\text{pH} = 1/2(pK_{a1} + pK_{a2}) = 6.32$$

(d) $HQ^+ + H_2O \rightleftharpoons H_2Q^{2+} + OH^- \qquad K_{b2} = 1.35 \times 10^{-10}$

At 75.00 mL HCl added, $V = 175$ mL, and

$$[HQ^+]_0 = [H_2Q^{2+}]_0 = \frac{2.50 \text{ mmol}}{175 \text{ mL}} = 0.0143 \text{ M}$$

$$\frac{(0.0143 + x)x}{0.0143 - x} = 1.35 \times 10^{-10}$$

$$x = [\text{OH}^-] = 1.35 \times 10^{-10} \text{ M}; \quad \text{pH} = 4.13$$

(e) $V = 199.90$ mL

$$[\text{H}_2\text{Q}^+]_0 = 0.500 \text{ M} \times \frac{99.9 \text{ mL}}{199.90 \text{ mL}} = 2.499 \times 10^{-2} \text{ M}$$

$$[\text{HQ}^+]_0 = 0.0500 \text{ M} \times \frac{0.10 \text{ mL}}{199.90 \text{ mL}} = 2.5 \times 10^{-5} \text{ M}$$

$$\text{H}_2\text{Q}^{2+} + \text{H}_2\text{O} \rightleftharpoons \text{HQ}^+ + \text{H}_3\text{O}^+$$

$$\frac{(2.5 \times 10^{-5} + x)x}{2.499 \times 10^{-2} - x} = \frac{K_\text{w}}{K_\text{b2}} = 7.41 \times 10^{-5}$$

Solving with the quadratic formula gives

$$x = [\text{H}_3\text{O}^+] = 1.31 \times 10^{-3} \qquad \text{pH} = 2.88$$

(f) When 100.00 mL of HCl has been added

$$[\text{H}_2\text{Q}^{2+}] = 0.0250 - x$$

$$\text{H}_2\text{Q}^{2+} + \text{H}_2\text{O} \rightleftharpoons \text{HQ}^+ + \text{H}_3\text{O}^+$$

$$\frac{x^2}{0.0250 - x} = \frac{1.00 \times 10^{-14}}{1.35 \times 10^{-10}} = 7.41 \times 10^{-5}$$

$$x = 1.32 \times 10^{-3} \text{ M} \qquad \text{pH} = 2.88$$

(g) When 105.00 mL of 0.1000 M HCl has been added,

$$[\text{H}_3\text{O}^+]_0 = \frac{5.00 \text{ mL} \times 0.1000}{205.00 \text{ mL}} = 2.44 \times 10^{-3} \text{ M}$$

$$[\text{H}_3\text{O}^+] = 2.44 \times 10^{-3} + x; \quad [\text{HQ}^+] = x; \quad [\text{H}_2\text{Q}] = 0.0500 \left(\frac{100.00}{205.00} \right) - x$$

$$\frac{x(2.44 \times 10^{-3} + x)}{2.44 \times 10^{-2} - x} = 7.41 \times 10^{-5}$$

$$x = 5.8 \times 10^{-4}; \quad [\text{H}_3\text{O}^+] = 3.02 \times 10^{-3}; \quad \text{pH} = 2.52$$

10-74 (a) The procedure is exactly the same as in problem 10-73. The $\Delta H°$ for the autoionization of water per mole of the reaction written in the problem is $+55.94$ kJ mol^{-1}.

(b) At 0° C, $\Delta G° = -RT \ln K = -(8.315)(273.15)\ln(1.139\times10^{-15}) = 78,150$ J mol^{-1}
$\Delta S° = (\Delta H° - \Delta G°)/T = (55,940 - 78,150$ J mol$^{-1})/273.15$ K $= -81.31$ J K^{-1} mol^{-1}

(c) The pH of water is exactly 7.00 when K_w equals exactly 1.00×10^{-14}. Substituting in the expression

$$\ln K = -\frac{\Delta H°}{RT} + \frac{\Delta S°}{R}$$

gives

$$\ln(1.00 \times 10^{-14}) = -\frac{55.94 \times 10^3 \text{ J mol}^{-1}}{8.315 \text{ J mol}^{-1} \text{ K}^{-1}}\left(\frac{1}{T}\right) + \frac{-81.31 \text{ J K}^{-1}}{8.315 \text{ J mol}^{-1} \text{ K}^{-1}}$$

Solving for T gives the desired temperature as 299.6 K, close to 298.15, the standard-state temperature.

10-76 The correct answer is (c). The point in this problem is that the solution of $Ca(OH)_2$ is so dilute that it is essentially indistinguishable from pure water. Hence the pH of the solution is the pH of water, which is 7.00 (assuming 25°C).

10-78 The equation given in the problem is the sum of the chemical equations for the second and third acid ionizations of phosphoric acid added to two times the reverse of the equation for the second ionization of carbonic acid. Hence, the desired equilibrium constant is

$$K = \frac{K_{a2,H_3PO_4} K_{a3,H_3PO_4}}{(K_{a2,H_2CO_3})^2} = \frac{(6.23 \times 10^{-8})(2.2 \times 10^{-13})}{(4.8 \times 10^{-11})^2} = 5.9$$

10-80 The solution of the weak acid, acetic acid, and the solution of the strong acid, hydrochloric acid, have the same pH, according to the color of the indicator. Because acetic acid is a weak acid, its concentration in the solution must be greater than the concentration of the strong acid to bring the pH to this particular value. The solution of acetic acid contains more moles of acid, so it can neutralize more moles of base.

10-82 HF is so much stronger as an acid than HClO (its K_a is 6.6×10^{-4} compared to 3.0×10^{-8}) that it determines the pH. The initial concentration is $[HF]_0 = 0.23$ mol/3.60 L $= 0.0639$ M. Set $x = [H_3O^+] = [F^-]$. Substituting in the equilibrium expression gives

$$\frac{x^2}{0.0639 - x} = K_a = 6.6 \times 10^{-4}$$

$$x = 6.17 \times 10^{-3} = [H_3O^+] = [F^-]$$

$$[HF] = 0.0639 - x = 0.058 \text{ M}$$

$$pH = 2.21$$

At this pH, almost all the HClO will be in the acid form. Its concentration is

$$[HClO] = \frac{0.57 \text{ mol}}{3.60 \text{ L}} = 0.16 \text{ M}$$

$$[ClO^-] = \frac{K_a [HClO]}{[H_3O^+]} = 7.7 \times 10^{-7} \text{ M}$$

10-84 $NH_4^+ + CN^- \rightleftharpoons NH_3 + HCN$

$$\frac{[NH_3][HCN]}{[NH_4^+][CN^-]} = \frac{5.6 \times 10^{-10}}{6.17 \times 10^{-10}} = 0.9076$$

$$[NH_3] = [HCN] = x$$

$$[NH_4^+] = [CN^-] = 0.100 - x$$

In writing this, we assumed $[H_3O^+]$, $[OH^-] << [NH_3]$, $[HCN]$.

$$\frac{x^2}{(0.100 - x)^2} = 0.9076$$

$$\frac{x}{0.100 - x} = 0.9527$$

$$x = 0.0488$$

$$[H_3O^+] = \frac{[NH_4^+]}{[NH_3]} K_a = \frac{0.100 - 0.0488}{0.0488}(5.6 \times 10^{-10}) = 5.9 \times 10^{-10}$$

(Note: the same result would come from using the HCN / CN^- equilibrium with K_a (HCN)).

10-86 At pH 10.00, $[H_3O^+] = 1.0 \times 10^{-10}$, and

$$\frac{CO_3^{2-}}{HCO_3^-} = \frac{K_a}{[H_3O^+]} = \frac{4.8 \times 10^{-11}}{1.0 \times 10^{-10}} = 0.48$$

Moles $CO_3^{2-} = 0.48$(moles HCO_3^-) $= 0.48x$

$$(0.48x \text{ mol})(105.99 \text{ g mol}^{-1}) + (x \text{ mol})(84.01 \text{ g mol}^{-1}) = 10.0$$

$$x = 7.41 \times 10^{-2} \text{ mol NaHCO}_3$$

mass $NaHCO_3 = 6.23$ g; mass $Na_2CO_3 = 3.77$ g

10-88 The pH shoots up suddenly upon the addition of 1 more drop of sodium hydroxide solution. This is the indication of an end-point; thus $4.71 + 0.01 = 4.72$ mL of 0.0410 M NaOH is just enough to bring the titration to the first equivalence point. At that point, the chemical amount of base that has been added equals the chemical amount of phosphoric acid that was present. The chemical amount of base is $(0.0410 \text{ mol L}^{-1} \times 0.00472 \text{ L})$. This is therefore also the chemical amount of acid. The original concentration of the phosphoric acid is this quantity divided by 0.05000 L (50.00 mL) or 3.87×10^{-3} M.

10-90 The vinegar under analysis contains around 5 g of acetic acid per 100 g. Acetic acid is CH_3COOH ($\mathcal{M} = 60.05$ g mol^{-1}). If the density of the vinegar is 1.0 g cm^{-3}, then Anne Dalton proposes to titrate 50.00 mL sample of about 0.83 M acetic acid. Using 1.000 M NaOH she can expect to need around 42 mL of base for each titration. If she detects the equivalence point to within ± 0.02 mL, then her precision is $(\pm 0.02/42) \times 100 \approx 0.05\%$.

Charlie Cannizzaro's method of analysis would give an uncertainty of ± 0.01 in the pH and therefore an uncertainty of $10^{\pm 0.01}$ in the concentration of H_3O^+. Because 10 raised to the 0.01 power is 1.023, this means that Charlie's result for the hydronium ion concentration would have $\pm 2.3\%$ uncertainty. There is an equal uncertainty in $[CH_3COO^-]$, which is equal to $[H_3O^+]$. Using the K_a expression to solve for the acetic acid concentration gives a total uncertainty of about 4.6 %. Anne Dalton's method is more precise. Its drawback is that it is slower and requires more skill.

10-92 The exhalation of the $CO_2(g)$ through the shell of the egg should lead to a rise in the pH of the contents—a weak acid is being lost from the system.

10-94 The formula of the hydrogen tartrate anion is $HC_4H_4O_6^-$. The dissolution of potassium hydrogen tartrate gives this ion along with potassium ions. Hydrogen tartrate ion acts as a base in the reaction:

$$HC_4H_4O_6^-(aq) + H_3O^+(aq) \rightleftharpoons H_2C_4H_4O_6(aq) + H_2O(l)$$

to counteract the effect of an added acid and as an acid in the reaction:

$$HC_4H_4O_6^-(aq) + OH^-(aq) \rightleftharpoons C_4H_4O_6^{2-}(aq) + H_2O(l)$$

to counteract the effect of an added base.

10-96 The equilibrium constant K_w for the autoionization of water increases with increasing temperature. By Le Chatelier's principle, this reaction must be endothermic.

10-98 (a)

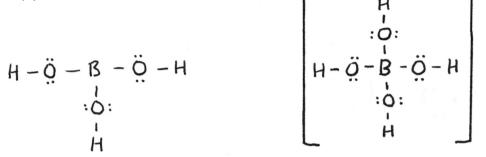

B(OH)$_3$ is a Lewis acid that accepts an electron pair from OH$^-$ to become the Lewis base, B(OH)$_4^-$.

(b) Let $x = [\text{B(OH)}_4^-] - [\text{H}_3\text{O}^+]$

$$\frac{x^2}{0.20 - x} = K_a = 5.8 \times 10^{-10}$$

$$x = 1.1 \times 10^{-5} = [\text{H}_3\text{O}^+]$$

$$\text{pH} = 4.97$$

10-100 From Henry's law, the mole fraction of CO$_2$ in solution is

$$X_{\text{CO}_2} = \frac{P_{\text{CO}_2}}{k} = \frac{0.833 \text{ atm}}{1.8 \times 10^3 \text{ atm}} = 4.63 \times 10^{-4}$$

One liter of water contains 55.5 mol water. The chemical amount of dissolved CO$_2$ is

$$\frac{n_{\text{CO}_2}}{n_{\text{CO}_2} + 55.5 \text{ mol}} = X_{\text{CO}_2} = 4.63 \times 10^{-4}$$

Solving gives $n_{\text{CO}_2} = 0.0257$ mol, so $[\text{CO}_2]_0 = 0.0257$ M $= [\text{H}_2\text{CO}_3]_0$. The pH for the diprotic acid is determined by the first acid dissociation. Let

$$x = [\text{HCO}_3^-] = [\text{H}_3\text{O}^+]$$

Then

$$\frac{x^2}{0.0257 - x} = K_{a1} = 4.3 \times 10^{-7}$$

$$x = 1.05 \times 10^{-4} = [\text{H}_3\text{O}^+]$$

$$\text{pH} = 3.98$$

Chapter 11

Heterogeneous Equilibria

11-2

$$\text{(a)} \quad \frac{1}{P_{C_2H_2}^3 P_{H_2}^3} = K \qquad \text{(b)} \quad \frac{P_{CO}^2}{P_{CO_2}} = K$$

$$\text{(c)} \quad \frac{P_{HF}^4 P_{CO_2}}{P_{CF_4}} = K \qquad \text{(d)} \quad P_{F_2} = K$$

11-4

$$\text{(a)} \quad \frac{[OH^-]^8 [I_2]^3}{[I^-]^6 [MnO_4^-]^2} = K \quad \text{(b)} \quad \frac{[I_2]}{[Cu^{2+}]^2 [I^-]^4} = K \quad \text{(c)} \quad \frac{[H_3O^+]^2}{P_{O_2}^{1/2} [Sn^{2+}]} = K$$

11-6 (a) When $P_{N_2O_4}$ is graphed against $P_{NO_2}^2$, they define a line that is close to straight. The slope of this line is the numerical value of K for the reaction $2\,NO_2 \rightleftharpoons N_2O_4$

(b) The seven pairs of equilibrium concentrations give seven different values of K. Their mean is 28.4. The graph can also be fitted by linear least-squares.

11-8 (a) Both gases come only from the volatilization of the ammonium carbamate; hence their partial pressures are related: $P_{NH_3} = 2P_{CO_2}$. Also, the sum of the two partial pressures is known: $P_{NH_3} + P_{CO_2} = 0.115$ atm. These two equations in two unknowns are easily solved to give

$$P_{NH_3} = 0.0767 \text{ atm} \quad \text{and} \quad P_{CO_2} = 0.0383 \text{ atm}$$

(b) Inserting the two partial pressures in the equilibrium expression gives the value of the equilibrium constant:

$$P_{NH_3}^2 P_{CO_2} = (0.0767)^2 (0.0383) = 2.25 \times 10^{-4} = K$$

11-10 (a)

$$\frac{(P_{HI})^2}{P_{H_2}} = 0.345$$

$$P_{HI}^2 = 0.345 P_{H_2} = 0.345(1.00) = 0.345$$

$$P_{HI} = 0.587 \text{ atm}$$

(b)

$$H_2(g) \quad + \quad I_2(s) \quad \rightleftharpoons \quad 2\,HI(g)$$
$$4.00 - x \qquad\qquad\qquad\qquad 2x$$

$$\frac{(2x)^2}{4.00 - x} = K = 0.345$$

$$4x^2 + 0.345x - 1.38 = 0$$

$$x = \frac{-0.345 \pm \sqrt{(0.345)^2 + 4(4)(1.38)}}{8} = 0.546$$

$$P_{H_2} = 4.00 - x = 3.45 \text{ atm}$$

$$P_{HI} = 2x = 1.09 \text{ atm}$$

11-12 The reaction quotient has the form

$$Q = \frac{P_{HI}^2}{P_{H_2S} P_{I_2}}$$

(a) Putting the values given in the problem into the expression gives $Q = 0$; only reactants are present. Some sulfur must be produced to reach equilibrium.

(b) For this set of initial conditions $Q = 3.5 \times 10^3$, which exceeds K. Solid sulfur is consumed by the reaction coming to equilibrium.

11-14 (a) The partial pressures of the two product gases (H_2O and CO_2) must equal each other and add up to 1.648 atm; hence they both equal 0.824 atm. Inserting the two partial pressures in the equilibrium expression gives the value of the equilibrium constant:

$$P_{H_2O} P_{CO_2} = (0.824)(0.824) = 0.679 = K$$

(b) $P_{H_2O} = K/P_{CO_2} = 0.697 \,/\, 0.800 = 0.849 \text{ atm}$

11-16 The equilibrium concentration of iodine in the aqueous layer is given as 4.16×10^{-5} M; the volume of this layer is not given. Suppose that it is V L. Then the chemical amount of I_2 in the aqueous layer at equilibrium is $(4.16 \times 10^{-2}V)$ mol. Before equilibrium (before the mixture was shaken), the amount of I_2 in the aqueous layer was much more—$(2.50 \times 10^{-2}V)$ mol. The amount of I_2 transferred to the CS_2 layer in attaining the partition equilibrium was

$$((2.50 \times 10^{-2}) - (4.16 \times 10^{-5})) \text{ mol L}^{-2} \times V \text{ L} \approx (2.50 \times 10^{-2})V \text{ mol}$$

The concentration of the iodine in the CS_2 is this chemical amount divided by the volume of the CS_2, which equals V L. It is therefore 2.50×10^{-2} mol L^{-1}. The partition equilibrium constant is

$$\frac{[I_2]_{(CS_2)}}{[I_2]_{(aq)}} = \frac{2.50 \times 10^{-2}}{4.16 \times 10^{-5}} = 600 = K$$

11-18 (a) The equilibrium constant for the dissolution of the citric acid according to the equation given in the problem is, in simple theory, just the concentration of the dissolved citric acid. For citric acid $\mathcal{M}$ is 192.1 g mol^{-1}, so the saturated aqueous solution contains 6.77 mol L^{-1} and $K = 6.8$. This assumes that there is no ionization of the citric acid in water and also that this concentrated solution behaves ideally. Neither assumption is very defensible. The K for the dissolution in ether is, by similar reasoning, 0.11.

(b) The transfer of citric acid from water into ether equals the reverse of the dissolution of citric acid in water added to the dissolution of citric acid in ether. The partition coefficient K is therefore $0.11/6.8 = 0.017$.

11-20 Even "bone-dry" residues can contain water of hydration, as indeed do dry bones. The 1.00 g sample of $MgSO_4$ amounts to 8.308×10^{-3} mol. It picks up 0.898 g of water when the solution is evaporated in one temperature range but only 0.150 g of water when the solution is evaporated in the second temperature range. These amounts are 0.0498 and 0.00833 mol. The chemical amount of water associated in the solid is 6.00 times the chemical amount of the magnesium sulfate in the first case and 1.00 times the chemical amount of the magnesium sulfate in the second case. The two formulas therefore are $MgSO_4 \cdot 6H_2O$ and $MgSO_4 \cdot H_2O$.

11-22 The 255 g of $AgNO_3$ is 1.50 mol of $AgNO_3$. The dissolution of this much silver nitrate in 100 g of water is just like the dissolution of 15.0 mol of silver nitrate in 1.00 kg of water. The graph (in Fig. 11.7) shows that 1.00 kg of water at 95°C easily accommodates 15.0 mol of $AgNO_3$, but that cooling the solution to about 30°C would cause $AgNO_3(s)$ to tend to precipitate.

11-24 The equation and K_{sp} expression for the dissolution are

$$Pb_3(SbO_4)_2(s) \rightleftharpoons 3\,Pb^{2+}(aq) + 2\,SbO_4^{3-}(aq) \qquad K_{sp} = [Pb^{2+}]^3[SbO_4^{3-}]^2$$

11-26 The dissolution of TlSCN gives Tl^+ and SCN^- ions; $K_{sp} = [Tl^+][SCN^-] = 1.82 \times 10^{-4}$. If neither of the ions react significantly to form other species, then the concentrations of the two are equal at equilibrium:

$$[Tl^+] = [SCN^-] = \sqrt{1.82 \times 10^{-4}} = 1.35 \times 10^{-2} \text{ mol L}^{-1}$$

The molar mass of TlSCN is 262.47 g mol^{-1} so 3.54 g of it dissolves per liter, or 0.354 g per 100 mL.

11-28 The reaction is

$$(NH_4)_2(PtCl_6)(s) \rightleftharpoons 2\,NH_4^+(aq) + PtCl_6^{2-}(aq)$$

$$[NH_4^+]^2[PtCl_6^{2-}] = K_{sp} = 5.6 \times 10^{-6}$$

$$\text{Let } S = [PtCl_6^{2-}], \text{ so that } [NH_4^+] = 2S$$

$$(2S)^2 S = 5.6 \times 10^{-6}$$

$$S = 0.011 \text{ mol L}^{-1}$$

$$(0.011 \text{ mol L}^{-1})(443.87 \text{ g mol}^{-1}) = 5.0 \text{ g L}^{-1}$$

11-30 The dissolution-precipitation equilibrium is $Hg_2Cl_2(s) \rightleftharpoons Hg_2^{2+}(aq) + 2\,Cl^-(aq)$. From the chemical equation, one mole of mercury(I) ions forms in solution for every one mole of solid that dissolves. Let S equal the concentration of mercury(I) present at equilibrium. Then the concentration of chloride is $2S$. The solubility-product expression is $K_{sp} = [Hg_2^{2+}][Cl^-]^2 = S(2S)^2 = 4S^3$. Solving for S gives a concentration of 8×10^{-7} M for the mercury(I) ion and 2×10^{-6} M for the chloride ion.

11-32 The molar mass of lead(II) iodate is 557.0 g mol^{-1}. Its solubility is

$$S = \frac{0.00896 \text{ g}}{0.400 \text{ L}} \times \frac{1 \text{ mol}}{557.0 \text{ g}} = 4.02 \times 10^{-5} \text{ M}$$

But, from the equation for the dissolution of the salt, $K_{sp} = S(2S)^2$. Substituting S gives $K_{sp} = 2.6 \times 10^{-13}$.

11-34 The molar mass of the silver dichromate is 431.72 g mol^{-1}, so the solubility of the salt is 1.31×10^{-4} mol L^{-1}. The value of K_{sp} is obtained by substitution in the K_{sp} expression. It is 9.0×10^{-12}.

11-36 The 0.090 g of PbI_2 ($\mathcal{M} = 461.0$ g mol^{-1}) is 1.95×10^{-4} mol. The concentration of Pb^{2+} in the hot solution is therefore 1.95×10^{-4} M, and that of I$^-$ is twice this, 3.90×10^{-4} M. These molarities hardly change as the solution is cooled, but the K_{sp} of the dissolution-precipitation equilibrium does change. At 25°C, it becomes equal to 1.4×10^{-8} (see Table 11.3). Does the reaction quotient Q exceed K_{sp}? It equals $(1.95 \times 10^{-4})(3.90 \times 10^{-4})^2 = 3.0 \times 10^{-11}$, which is <u>smaller</u> than K_{sp}. No precipitate forms.

11-38 The calcium chloride and sodium fluoride solutions dilute each other as they are mixed. The concentration of Ca^{2+} is 2/3 of 0.0010 M immediately after mixing, and the concentration of F$^-$ is 1/3 of 6.0×10^{-5} M. Substituting these concentrations in the K_{sp} expression for CaF_2 gives a Q of about 2.7×10^{-13}. Because this is less than K_{sp}, no precipitate forms.

11-40 The volume of the combined solutions is 8.10 L. After mixing but before any reaction occurs, the concentrations of the reacting ions are

$$[Ag^+] = \left(\frac{1.50}{8.10}\right)(0.080) = 0.0148 \text{ M} \quad \text{and} \quad [I^-] = \left(\frac{6.60}{8.10}\right)(0.10) = 0.0815 \text{ M}$$

The two ions react in a 1:1 proportion, so Ag$^+$ ion is the limiting reactant. At equilibrium, almost all of the Ag$^+$ is removed from solution as AgI(s). The remaining concentration of I$^-$(aq) is $0.0815 - 0.0148 = 0.0667$ M. Inserting this value into the K_{sp} expression gives

$$[Ag^+][I^-] = 1.5 \times 10^{-16} = [Ag^+][0.0667]$$

Solving gives an equilibrium concentration of Ag$^+$ of 2.25×10^{-15} M. Only a fraction 1.5×10^{-13} of the silver ions originally in solution remain.

11-42 The reaction responsible for the precipitate is

$$Sr^{2+}(aq) + 2\,F^-(aq) \rightleftharpoons SrF_2(s)$$

The chemical amount of F$^-$ ion is 4.00 mmol, and the chemical amount of Sr^{2+} is 3.20 mmol. Imagine that all of the F$^-$ ion reacts (it is the limiting reactant). Then 1.20 mmol of Sr^{2+} remains. The concentration of this remaining Sr^{2+} is 1.20 mmol/120 mL $= 0.010$ M. Now, suppose that the $SrF_2(s)$ starts to redissolve. Some Sr^{2+} comes into solution but the amount is negligible compared to the amount of unreacted excess Sr^{2+}. The dissolution of SrF_2 is however the <u>only</u> source of F$^-$(aq). The dissolution proceeds until the K_{sp} relationship is satisfied:

$$K_{sp} = [Sr^{2+}][F^-]^2 = 2.8 \times 10^{-9}$$

Inserting $[Sr^{2+}] = 0.010$ M gives 5.3×10^{-4} M as the equilibrium concentration of the fluoride ion.

11-44 The common-ion effect greatly reduces the solubility of the AgCl, which is only very slightly soluble in pure water in the first place. Let S represent this solubility. Then, $[Cl^-] = (0.150 + S)$ M and $[Ag^+] = S$ M. Substituting in the K_{sp} expression gives

$$K_{sp} = [Ag^-][Cl^-] = S(0.150 + S) = 1.6 \times 10^{-10}$$

Solving for S gives the molar solubility of AgCl under these circumstances. It is 1.07×10^{-9} M. This means that 1.07×10^{-10} mol of AgCl dissolves in 100 mL of the 0.150 M NaCl. Taking M of AgCl as 143.3 g mol^{-1} gives the gram-solubility as 1.5×10^{-8} g per 100 mL.

11-46 (a) Let S equal the molar solubility of the silver arsenate. From the equilibrium law for the dissolution-precipitation reaction

$$K_{sp} = 1.0 \times 10^{-22} = [Ag^+]^3[AsO_4^{3-}] = (3S)^3 S$$

Solving gives $S = 1.4 \times 10^{-6}$ M

(b) Now, use $[Ag^+] = 0.100$ M in the equilibrium expression and let the concentration of AsO_4^{3-} equal S. Substituting and solving gives $S = 1.0 \times 10^{-19}$ mol L^{-1}.

11-48 Let S = solubility of BaF_2 in pure water.

$$S(2S)^2 = K_{sp} = 1.7 \times 10^{-6}$$

$$S = 7.5 \times 10^{-3} = [Ba^{2+}]$$

After the concentration of Ba^{2+} is reduced to 1.0% of this,

$$[Ba^{2+}] = 7.5 \times 10^{-5} \text{ M}$$

The fluoride concentration is then

$$[F^-] = \sqrt{K_{sp}/[Ba^{2+}]} = 0.15 \text{ M}$$

11-50 (a) Assume that neither the Mg^{2+} ion nor the OH^- ion from the dissolution of the $Mg(OH)_2$ interacts further in the solution. Then, if S is the solubility of the $Mg(OH)_2$, the concentration of Mg^{2+} ion is S and the concentration of OH^- ion is $2S$. At equilibrium,

$$K_{sp} = 1.2 \times 10^{-11} = S(2S)^2 \text{ from which } S = 1.4 \times 10^{-4} \text{ mol L}^{-1}$$

(b) If the pH is buffered at a value of 9, then the $[OH^-]$ is being held at 10^{-5} M. This concentration is *less* than what forms from the dissolution of the

$Mg(OH)_2$ in pure water. After the dissolution of $Mg(OH)_2$ comes to equilibrium, the concentration of OH^- ion remains at 10^{-5} M because of the action of the buffer. Let S again represent the solubility of the salt. Then:

$$K_{sp} = 1.2 \times 10^{-11} = S(10^{-5})^2$$

and the solubility is 0.12 mol L^{-1}.

11-52 (a) Solubility will increase as CO_3^{2-} reacts with H_3O^+ to give HCO_3^-.

(b) Solubility will show little change because HBr is a strong acid.

(c) Solubility will increase as HS^- (from MnS) reacts with H_3O^+ to give H_2S.

11-54 (a) If the F^- ion concentration exceeds 8.8×10^{-6} M, then CaF_2 precipitates. If it exceeds 4.1×10^{-3} M, BaF_2 precipitates. These two concentrations are calculated by substitution into the two K_{sp} expressions. If we maintain the $[F^-]$ below 4.1×10^{-3} M, only the CaF_2 precipitates.

(b) If the $[F^-]$ is held at 4.1×10^{-3} M then the concentration of Ca^{2+} in equilibrium with it and the $CaF_2(s)$ is only 2.3×10^{-6} M. This is a fraction 4.6×10^{-6} of the 0.50 M concentration originally present.

11-56 The $BaSO_4$ is more insoluble than the $CaSO_4$. Substitution of the given $[Ca^{2+}]$ concentration into the K_{sp} expression for $CaSO_4$ gives the concentration of SO_4^{2-} at which $CaSO_4$ just starts to precipitate. It is 3.0×10^{-4} M. At this concentration of sulfate ion, $[Ba^{2+}]$ is 3.7×10^{-7} M, according to the K_{sp} expression for $BaSO_4$.

11-58 From Table 11.4,

$$CdS(s) + H_2O(l) \rightleftharpoons Cd^{2+}(aq) + OH^-(aq) + HS^-(aq) \qquad K = 7 \times 10^{-28}$$

Add to this the equilibria

$$HS^-(aq) + H_3O^+(aq) \rightleftharpoons H_2S(aq) + H_2O(l) \qquad K = \frac{1}{K_a}$$

$$H_3O^+(aq) + OH^-(aq) \rightleftharpoons 2\,H_2O(l) \qquad K = \frac{1}{K_w}$$

This gives

$$CdS(s) + 2\,H_2O^+(aq) \rightleftharpoons Cd^{2+}(aq) + H_2S(aq) + 2\,H_2O(l)$$

$$\frac{[Cd^{2+}][H_2S]}{[H_3O^+]^2} = \frac{K}{K_a K_w} = \frac{7 \times 10^{-28}}{(9.1 \times 10^{-8})(1.0 \times 10^{-14})} = 7.7 \times 10^{-7}$$

Inserting $[H_2S] = 0.10$ M and $[H_3O^+] = 1.0 \times 10^{-3}$ M and solving for $[Cd^{2+}]$ gives

$$[Cd^{2+}] = 8 \times 10^{-12} \text{ M}$$

11-60 For Mn^{2+} to remain in solution, we must have (following the procedure in problem 11-58)

$$\frac{[Mn^{2+}][H_2S]}{[H_3O^+]^2} < \frac{K}{K_a K_w} = \frac{3 \times 10^{-14}}{(9.1 \times 10^{-8})(1.0 \times 10^{-14})} = 3.3 \times 10^7$$

Inserting $[Mn^{2+}] = 0.050$ M and $[H_2S] = 0.10$ gives

$$[H_3O^+]^2 > \frac{[Mn^{2+}][H_2S]}{3.3 \times 10^7} = \frac{(0.050)(0.10)}{3.3 \times 10^7} = 1.5 \times 10^{-10}$$

$$[H_3O^+] > 1.2 \times 10^{-5}; \qquad \text{pH} < 4.9$$

Now rewrite the expression for CdS dissolution from problem 11-58:

$$[Cd^{2+}] = \frac{[H_3O^+]^2}{[H_2S]}(7.7 \times 10^{-7}) = \frac{(1.2 \times 10^{-5})^2}{0.10}(7.7 \times 10^{-7}) = 1 \times 10^{-15} \text{ M}$$

11-62 The term "formation constant" refers to the chemical equilibrium for the formation of the complex ion:

$$Tl^{3+}(aq) + 4\,Cl^-(aq) \rightleftharpoons TlCl_4^-(aq)$$

The large K_f means that the complex is heavily favored at equilibrium. Imagine that the Tl^{3+} from the $Tl(NO_3)_3$ reacts to completion with the available Cl^- from the NaCl to form the complex. The Tl^{3+} is in excess, so the concentration of Cl^- would then be zero, the concentration of Tl^{3+} would be 0.025 M, and the concentration of $TlCl_4^-$ would be 0.125 M. Next, imagine that equilibrium is attained by subsequent partial break-up of the complex. Because K_f is so large, the equilibrium concentrations do not differ significantly from the values just listed. No additional calculations are needed unless it is desired to verify that the equilibrium concentration of Cl^- is indeed small.

11-64 Let us abbreviate the 18-crown-6 as "crown." We have the two equilibria:
$K^+ + \text{crown} \rightleftharpoons \text{K-crown}^+ \qquad K = 1.41 \times 10^6$
$Cs^+ + \text{crown} \rightleftharpoons \text{Cs-crown}^+ \qquad K = 2.75 \times 10^4$

Both equilibrium constants for the formation of crown complexes are large, so both equilibria lie far to the right. The crown compound is in excess in the solution. Thus, effectively all of the K^+ and Cs^+ ion are consumed leaving

$0.30 - 0.020 - 0.020 = 0.26$ mol L^{-1} of crown and forming 0.020 mol L^{-1} of K-crown$^+$ and 0.020 mol L^{-1} of Cs-crown$^+$. We construct the appropriate equilibrium expressions and substitute the concentrations just listed:

$$K = \frac{[\text{Kcrown}^+]}{[\text{K}^+][\text{crown}]} = 1.41 \times 10^6$$

$$1.41 \times 10^6 = \frac{0.020}{[\text{K}^+]0.26}; \quad [\text{K}^+] = 5.5 \times 10^{-8} \text{ M}$$

A similar computation on the Cs$^+$ equilibrium gives $[\text{Cs}^+] = 2.8 \times 10^{-6}$ M.

11-66

$$\text{AgCl}(s) \rightleftharpoons \text{Ag}^+(aq) + \text{Cl}^-(aq) \qquad K_{\text{sp}} = 1.7 \times 10^{-10}$$

$$\text{Ag}^+(aq) + 2\,\text{NH}_3(aq) \rightleftharpoons \text{Ag(NH}_3)_2^+(aq) \qquad K_{\text{f}} = 1.7 \times 10^7$$

Adding the second equation to the first gives

$$\text{AgCl}(s) + 2\,\text{NH}_3(aq) \rightleftharpoons \text{Ag(NH}_3)_2^+(aq) + \text{Cl}^-(aq)$$

$$K = K_{\text{sp}}K_{\text{f}} = 2.9 \times 10^{-3}$$

$$[\text{NH}_3] = 1.0 - 2x$$

$$[\text{Ag(NH}_3)_2^+] = x = [\text{Cl}^-]$$

$$\frac{x^2}{(1.0 - 2x)^2} = 2.9 \times 10^{-3}$$

$$\frac{x}{1.0 - 2x} = 0.054$$

$$x = 0.049 = [\text{Ag(NH}_3)_2^+]$$

0.049 mol AgCl dissolves per liter, or $(0.049 \text{ mol})(143.3 \text{ g mol}^{-1}) = 7.0$ g.

11-68 The solution of FeCl$_3$ will be acidic. The Fe^{3+} ion interacts with water

$$\text{Fe}^{3+}(aq) + 2\,\text{H}_2\text{O}(l) \rightleftharpoons \text{FeOH}^{2+}(aq) + \text{H}_3\text{O}^+(aq)$$

The reaction can also be represented:

$$\text{Fe(H}_2\text{O})_6^{3+}(aq) + \text{H}_2\text{O}(l) \rightleftharpoons \text{Fe(H}_2\text{O})_5\text{OH}^{2+}(aq) + \text{H}_3\text{O}^+(aq)$$

11-70 The reaction is

$$Fe(H_2O)_6^{2+}(aq) + H_2O(l) \rightleftharpoons Fe(H_2O)_5OH^+(aq) + H_3O^+(aq)$$

for which the K_a-expression is

$$K_a = 3 \times 10^{-6} = \frac{[H_3O^+][Fe(H_2O)_5OH^+]}{[Fe(H_2O)_6^{2+}]}$$

The concentrations of $Fe(H_2O)_5OH^+$ and H_3O^+ are equal. Represent them by x. Then

$$3 \times 10^{-6} = \frac{x^2}{0.10 - x}$$

and $x = [H_3O^+] = 5.5 \times 10^{-4}$ M. The corresponding pH is 3.3 which is less acidic that the pH 1.6 of a 0.1 M solution of $Fe(NO_3)_3$.

11-72 We treat the solution of $Ni(NO_3)_2$ as containing a weak acid that donates one hydrogen ion. The equation is

$$Ni(H_2O)_6^{2+}(aq) + H_2O(l) \rightleftharpoons Ni(H_2O)_5OH^+(aq) + H_3O^+(aq)$$

Assuming that the concentration of the $Ni(H_2O)_5OH^+$ ion equals that of the H_3O^+ ion, then both equal 1×10^{-5} M. Substituting in the K_a expression gives

$$K_a = \frac{(1 \times 10^{-5})^2}{(0.10 - 1 \times 10^{-5})} = 1 \times 10^{-9}$$

11-74 At pH 14.0, if we assume $Zn(OH)_2(s)$ is present, we have

$$[Zn^{2+}][OH^-]^2 = K_{sp} = 4.5 \times 10^{-17}$$

$$[Zn^{2+}] = \frac{4.5 \times 10^{-17}}{(1.0)^2} = 4.5 \times 10^{-17} \text{ M}$$

$$\frac{[Zn(OH)_4^{2-}]}{[Zn^{2+}][OH^-]^4} = K_f = 5 \times 10^{14}$$

$$[Zn(OH)_4^{2-}] = K_f[Zn^{2+}][OH^-]^4 = (5 \times 10^{14})(4.5 \times 10^{-17})(1.0)^4 = 2.2 \times 10^{-2}$$

$$[Zn(OH)_4^{2-}] + [Zn^{2+}] = 2.2 \times 10^{-2} + 4.5 \times 10^{-17} = 0.022 \text{ M} > 0.01 \text{ M}$$

so that there is a contradiction. Thus our assumption was false and no precipitate forms. We must therefore recalculate the concentrations using

$$\frac{[Zn^{2+}]}{[Zn(OH)_4^{2-}]} = \frac{1}{K_f[OH^-]^4} = 2 \times 10^{-15}$$

$$[\text{Zn(OH)}_4^{2-}] = 0.010 \text{ M and } [\text{Zn}^{2+}] = 2 \times 10^{-17} \text{ M}$$

In the second case,

$$[\text{Zn}^{2+}] = \frac{(4.5 \times 10^{-17})}{(0.10)^2} = 4.5 \times 10^{-15} \text{ M}$$

$$[\text{Zn(OH)}_4^{2-}] = (5 \times 10^{14})(4.5 \times 10^{-15})(0.1)^4 = 2.2 \times 10^{-4}$$

$$[\text{Zn(OH)}_4^{2-}] + [\text{Zn}^{2+}] = 2.2 \times 10^{-4} < 0.10 \text{ M}$$

so that in this case our assumption was correct and a precipitate forms.

11-76 The second equation in the problem is the first reversed and multiplied through by 2. Hence, $K_2 = 1/K_1^2$ where the subscripts refer to the equations in the order given in the problem.

11-78 Write the equilibrium law for the reaction of $\text{KOH}(s)$ with $\text{CO}_2(g)$ to give $\text{KHCO}_3(s)$. Two of the three compounds in the equilibrium are solids and do not appear in the expression, which is consequently quite simple:

$$\frac{1}{P_{\text{CO}_2}} = K$$

The K is given as 6×10^{15} so $P_{\text{CO}_2} = 2 \times 10^{-16}$ atm. The amounts of the two solids are immaterial except to assure that both solids are present at equilibrium.

11-80 (a) According to the balanced equation, one mole of fructose forms for every one mole of glucose that reacts. At equilibrium in the first experiment, [fructose] $= 0.1175$ M and [glucose] $= 0.2564 - 0.1175 = 0.1389$ M. The equilibrium constant is the ratio of these two numbers, which is 0.8459. In the second experiment, equilibrium is approached from the other direction. At equilibrium [fructose] $= 0.2666 - 0.1415 = 0.1251$ M and [glucose] $= 0.1415$ M. The K is 0.8841, which is very similar. (These are published experimental data). The average K is 0.865.
(b) Let us use the average equilibrium constant just calculated. If x is the fraction of glucose converted, then at equilibrium [fructose] $= xC_o$ where C_o is the initial concentration. Then [glucose] $= (1 - x)C_o$ and, $\frac{x}{1-x} = 0.865$ Solving for x gives 0.464, so 46.4% of the glucose is converted.

11-82 From problems 11-17 and 11-18:

$$K_1 = \frac{[\text{benzoic acid}]_{(ether)}}{[\text{benzoic acid}]_{(aq)}} = 330 \text{ and } K_2 = \frac{[\text{citric acid}]_{(ether)}}{[\text{citric acid}]_{(aq)}} = 1.69 \times 10^{-2}$$

Ether and water are immiscible; they form layers when mixed. The data mean that when the 50:50 mixture of citric acid and benzoic acid is treated with the mixture of ether and water, most of the benzoic acid ends up in the ether, and most of the citric acid ends up in the water.

Let x equal the mass of the benzoic acid present in the water at equilibrium. Then $1.00 - x$ is the mass of benzoic acid present in the ether. Similarly, let y be the mass of the citric acid present in the water at equilibrium. Then $1.00 - y$ is the mass of citric acid in the ether. The equilibrium laws for the partition of the two organic acids involve concentrations, not amounts. In this problem, the volumes of the two solvents are equal, so the amounts of the solutes in the two layers have the same ratios as their concentrations. Then:

$$\frac{1.00 - x}{x} = 330 \quad \text{and} \quad \frac{1.00 - y}{y} = 1.69 \times 10^{-2}$$

Solving the equation in x shows that there is 3.0×10^{-3} g of benzoic acid in the water at equilibrium and 0.9970 g of benzoic acid in the ether. From the equation in y, there is 0.9834 g of citric acid in the water at equilibrium and 0.0166 g of citric acid in the ether. The total mass of the solids in the ether is 1.0136 g, of which 0.9970 g is benzoic acid. The benzoic acid is therefore 98.4 % of the dissolved solids in the ether. The total mass of the solids in the water is 0.9864 g, of which 0.9834 g is citric acid. The citric acid is 99.7 % of the solid recovered from the water layer.

11-84 (a) In the van't Hoff equation

$$\ln \frac{K_2}{K_1} = \frac{-\Delta H^\circ}{R} \left(\frac{1}{T_2} - \frac{1}{T_1} \right)$$

the two K's are $K_1 = 4.5 \times 10^{-19}$ and $K_2 = 6.2 \times 10^{-12}$ and $T_1 = 1000$ K and $T_2 = 1200$ K. Substitution gives ΔH° equal to 8.2×10^5 J mol^{-1}. This means that the ΔH° of the reaction as written is 820 kJ.

(b) Another substitution of K's and T's in the van't Hoff equation gives a ΔH° equal to 16.4×10^5 J mol^{-1}. It follows that the ΔH° of the reaction written with doubled coefficients is 1640 kJ as thermodynamics requires.

11-86 If the peroxydisulfuryl difluoride (which we shall call the dimer) behaved ideally, then we would expect the pressure of the sample to increase by a factor of $383.15/373.15 = 1.027$ when the temperature was raised from 100° to 110°C. The breakdown of the dimer to FO_2SO (the monomer) would cause the pressure to increase even more than this because it creates two moles of gas from every one mole of gas that reacts. Yet, the pressure of the sample changes only by 1.027. We conclude that the proportion of FO_2SO in the sample must be small after the temperature increase, and even smaller before the temperature increase (higher temperature favors the dark-colored monomer). Although the proportion of the monomer remains small throughout the experiment, its concentration nevertheless doubles between 100 and 110° according to the change in the intensity of its color. With these points in mind, we write the equilibrium law for the reaction dimer $\rightleftharpoons$ 2 monomer. The form of the law is the same at the two temperatures. At 100°C it is:

$$K_{373} = \frac{P^2_{\text{monomer}}}{P_{\text{dimer}}} = \frac{[\text{monomer}]^2}{[\text{dimer}]} RT = \frac{[\text{monomer}]^2}{[\text{dimer}]}(373.15)\,R$$

where we have used the fact that $P = (n/V)RT$ for each component. At 110°C it is

$$K_{383} = \frac{P^2_{\text{monomer}}}{P_{\text{dimer}}} = \frac{[\text{monomer}]^2}{[\text{dimer}]}(383.15)\,R$$

We want to use the ratio of these two K's in the van't Hoff equation. Dividing the second equation by the first gives:

$$\frac{K_{383}}{K_{373}} = \left(\frac{[\text{monomer}]_{383}}{[\text{monomer}]_{373}}\right)^2 \left(\frac{[\text{dimer}]_{373}}{[\text{dimer}]_{383}}\right)\left(\frac{383.15}{373.15}\right)$$

The concentration of monomer doubles with the temperature increase so the first term in parentheses equals 4. The second term essentially equals 1 because the amount of the dimer in the container hardly changes. Thus:

$$\frac{K_{383}}{K_{373}} = (4) \times (1) \times \left(\frac{383.15}{373.15}\right) = 4.11$$

Substituting in the van't Hoff equation gives

$$\ln \frac{K_{383}}{K_{373}} = \ln 4.11 = \frac{-\Delta H^\circ}{R}\left(\frac{1}{383.15} - \frac{1}{373.15}\right)$$

and the molar ΔH° of the reaction is easily computed. For every mole of the reaction as it is written $\Delta H^\circ = 168$ kJ $= 1.7 \times 10^2$ kJ.

11-88 The chemical equation is

$$MgNH_4PO_4 \cdot 6H_2O(s) \rightleftharpoons Mg^{2+}(aq) + NH_4^+(aq) + PO_4^{3-}(aq) + 6\,H_2O(l)$$

The six moles of water of crystallization are included in the equation to maintain a formal balance. They join the rest of the water in the solution, however, and are not included in the following K_{sp} expression:

$$[Mg^{2+}][NH_4^+][PO_4^{3-}] = K_{sp} = 2.3 \times 10^{-13}$$

Once this compound is dissolved, because $PO_4^{3-}(aq)$ is a stronger base than NH_3, the reaction $NH_4^+(aq) + PO_4^{3-}(aq) \rightleftharpoons NH_3(aq) + HPO_4^{2-}(aq)$ will occur.

11-90 The dissolution reaction is represented:

$$AgClO_4(s) \rightleftharpoons Ag^+(aq) + ClO_4^-(aq)$$

The large concentration of ClO_4^- present in 60 percent perchloric acid solution reduces the solubility of the $AgClO_4$ (the common-ion effect).

11-92 We write the solubility-product expression

$$[Ca^{2+}][F^-]^2 = K_{sp} = 3.9 \times 10^{-11}$$

With $[Ca^{2+}] = 0.0020$ M, the largest $[F^-]$ that can be tolerated before CaF_2 tends to precipitate is 14×10^{-5} M. This exceeds the recommended $F^-(aq)$ concentration.

11-94 Two dissolution-precipitation equilibria take place simultaneously. They are:

$$BaSO_4(s) \rightleftharpoons Ba^{2+}(aq) + SO_4^{2-}(aq) \quad \text{and} \quad AgCl(s) \rightleftharpoons Ag^+(aq) + Cl^-(aq)$$

We mix solutions of Ag_2SO_4 and $BaCl_2$, both of which are dissociated into ions. The concentrations of the four ions after the mixing, but before any precipitation has occurred are $[Ba^{2+}] = 0.020$ M, $[Ag^+] = 0.020$ M, $[Cl^-] = 0.040$ M, and $[SO_4^{2-}] = 0.010$ M. Assume that the Ba^{2+} and the SO_4^{2-} react completely to form $BaSO_4(s)$. Then the concentration of the sulfate is zero, and the concentration of the barium ion, which is in excess in the precipitation reaction, is 0.010 M. Substituting this 0.010 M into the K_{sp} expression gives $[SO_4^{2-}] = 1.1 \times 10^{-8}$ M, which is indeed very close to zero. In the same way, Cl^- is in excess relative to Ag^+. At equilibrium, its concentration will be 0.020 M. The K_{sp} for AgCl then gives $[Ag^+] = 8.0 \times 10^{-9}$ M.

11-96 When CaO dissolves in water, the oxide ion reacts immediatly with water to furnish hydroxide ions. The chemical equation is

$$CaO(s) + H_2O(l) \rightleftharpoons Ca^{2+}(aq) + 2\ OH^-(aq)$$

for which the equilibrium constant expression is $K = [Ca^{2+}][OH^-]^2$.

The dissolution of $CaO(s)$ in acid is better represented as

$$CaO(s) + 2\,H_3O^+(aq) \rightleftharpoons Ca^{2+}(aq) + 3\,H_2O(l)$$

The equilibrium constant is then

$$\frac{[Ca^{2+}]}{[H_3O^+]^2} = \frac{K}{K_w^2}$$

11-98

$$Ag^+(aq) + Cl^-(aq) \rightarrow AgCl(s)$$

$$AgCl(s) + 2\,NH_3(aq) \rightarrow Ag(NH_3)_2^+(aq) + Cl^-(aq)$$

$$Ag(NH_3)_2^+(aq) + 2\,H_3O^+(aq) + Cl^-(aq) \rightarrow AgCl(s) + 2\,NH_4^+(aq) + 2\,H_2O(l)$$

11-100 (a) Suppose x mol per liter CaC_2O_4 dissolves. Then $[Ca^{2+}] = x$. The $C_2O_4^{2-}$ can react with $H_2C_2O_4$ according to

$$C_2O_4^{2-}(aq) + H_2C_2O_4(aq) \rightleftharpoons 2\,HC_2O_4^-(aq)$$

$$\frac{[HC_2O_4^-]^2}{[C_2O_4^{2-}][H_2C_2O_4]} = \frac{K_{a1}}{K_{a2}} = 920$$

Suppose that at equilibrium $[C_2O_4^{2-}] = y$. Then $x - y$ mol per liter has reacted, and

$$[HC_2O_4^-] = 2(x - y)$$

$$[H_2C_2O_4] = 1.0 - (x - y)$$

We have two simultaneous equilibria:

$$\frac{[2(x-y)]^2}{y[1 - x + y]} = \frac{4(x-y)^2}{y(1 - x + y)} = 920$$

and

$$[Ca^{2+}][C_2O_4^{2-}] = K_{sp} = xy = 2.6 \times 10^{-9}$$

$$y = 2.6 \times 10^{-9}/x$$

$$\frac{4(x-y)^2 x}{(2.6 \times 10^{-9})(1 - x + y)} = 920$$

$$\frac{x(x-y)^2}{1 - x + y} = 5.98 \times 10^{-7}$$

Assume $y \ll x$. Then $\frac{x^3}{1-x} = 5.98 \times 10^{-7}$.

Now assume $x \ll 1$. Then $x^3 = 5.98 \times 10^{-7}$.

$$x = 8.4 \times 10^{-3} \ll 1$$

$$y = \frac{2.6 \times 10^{-9}}{8.4 \times 10^{-3}} = 3.1 \times 10^{-7} \ll x$$

Solubility $= [Ca^{2+}] = 8.4 \times 10^{-3}$ M.

(b) $[Ca^{2+}] = [C_2O_4^{2-}] = x$

$x^2 = 2.6 \times 10^{-9}$

$x = 5.1 \times 10^{-5}$ M

(c) Because $C_2O_4^{2-}$ can react with $H_2C_2O_4$, its concentration will decrease. By Le Chatelier's principle, the system will respond by further dissolution of $C_2O_4^{2-}$, increasing the solubility over that in pure water.

11-102 This is equivalent to making a solution that is initially 0.100 M in $HgCl^+$. We make the correspondences with Section 10.8 by noting that in acid-base equilibria we consider dissociation but in complex ions we have formation, the reverse:

$$
\begin{array}{rcl}
HgCl_2 & \leftrightarrow & H_2CO_3 \\
HgCl^+ & \leftrightarrow & HCO_3^- \\
Hg^{2+} & \leftrightarrow & CO_3^{2-} \\
Cl^- & \leftrightarrow & H_3O^+ \\
NO_3^- & \leftrightarrow & Na^+ \\
1/K_2 & \leftrightarrow & K_{a1} \\
1/K_1 & \leftrightarrow & K_{a2}
\end{array}
$$

where K_1 and K_2 correspond to the formation of $HgCl^+$ from Hg^{2+} and of $HgCl_2$ from $HgCl^+$. Because nothing corresponds to the hydroxide ion, we can set the analog of K_w to zero. Substituting into the equation for $[H_3O^+]$ on page 351 gives

$$[Cl^-]^2 = \frac{[HgCl^+]/K_1K_2}{1/K_2 + [HgCl^+]}$$

If we now assume $[HgCl^+] \gg 1/K_2$ (to be checked later) we find

$$[Cl^-]^2 = 1/K_1K_2 = 1/(5.5 \times 10^6)(3 \times 10^6)$$

$$[Cl^-] = 2.46 \times 10^{-7} \text{ M}$$

We can write

$$\frac{[HgCl_2]}{[HgCl^+][Cl^-]} = K_2$$

$$[HgCl_2] = K_2[Cl^-][HgCl^+] = 0.739[HgCl^+]$$

and

$$\frac{[HgCl^+]}{[Hg^{2+}][Cl^-]} = K_1$$

$$[Hg^{2+}] = \frac{1}{K_1[Cl^-]}[HgCl^+] = 0.739[HgCl^+]$$

The fact that $[Hg^{2+}] = [HgCl_2]$ results from stoichiometry as well. Adding total concentrations of Hg-containing species gives

$$[HgCl^+] + 2(0.739)[HgCl^+] = 0.100$$

$$[HgCl^+] = 0.040 \text{ M} \gg 1/K_2$$

$$[Hg^{2+}] = [HgCl_2] = 0.030 \text{ M}$$

11-104 $Pt(OH)_6^{2-}(aq) + H_2O(l) \rightarrow Pt(H_2O)(OH)_5^-(aq) + OH^-(aq)$

11-106

$$\frac{1}{(P_{H_2O})^2} = K = 1.6 \times 10^3 \text{ at equilibrium}$$

$$P_{H_2O} = 0.025 \text{ atm at equilibrium}$$

At this temperature, the equilibrium vapor pressure of water is 0.03126 atm, so the borderline relative humidity is $0.025/0.03126 \times 100\% = 80\%$. Above this value Q < K and $CaSO_4$ is converted to the hydrate.

11-108 The $\Delta G°$ for the reaction

$$SrSO_4(s) \rightarrow Sr^{2+}(aq) + SO_4^{2-}(aq)$$

is

$$-559.48 - 744.53 - (-1340.9) = +36.89 \text{ kJ}$$

at 25°C. From this, the equilibrium constant, K_{sp}, can be calculated:

$$\Delta G° = -RT \ln K_{sp}$$

$$36,890 \text{ J mol}^{-1} = -(8.3145 \text{ J mol}^{-1} \text{ K}^{-1})(298.15 \text{ K}) \ln K_{sp}$$

$$K_{sp} = 3.4 \times 10^{-7}$$

The molar solubility S of $SrSO_4$ is approximately related to K_{sp} through

$$[Sr^{2+}][SO_4^{2-}] = S^2 = K_{sp}$$

so that $S = 5.9 \times 10^{-4} \text{ mol L}^{-1}$.

11-110 (a) The sublimation of $NbI_5(s)$ is described by

$$NbI_5(s) \rightleftharpoons NbI_5(g) \qquad K = P_{NbI_5}$$

where K depends on the temperature according to the equation:

$$\ln K = -\frac{\Delta H^\circ_{subl}}{RT} + \frac{\Delta S^\circ_{subl}}{R}$$

Substituting the equilibrium expression into this equation gives

$$\ln P_{NbI_5} = -\frac{\Delta H^\circ_{subl}}{RT} + \frac{\Delta S^\circ_{subl}}{R}$$

The natural logarithm of a number is 2.3026 times the common (base 10) logarithm. Hence:

$$2.3026 \log P_{NbI_5} = -\frac{\Delta H^\circ_{subl}}{RT} + \frac{\Delta S^\circ_{subl}}{R}$$

The problem gives an expression for the dependence of $\log P_{NbI_5}$ on T. Insert it on the left

$$2.3026 \left(\frac{-6762}{T} + 8.566\right) = -\frac{\Delta H^\circ_{subl}}{RT} + \frac{\Delta S^\circ_{subl}}{R}$$

Term-by-term comparison of the two sides of this equation shows that

$$\Delta H^\circ_{subl} = 2.3026(6762R) \quad \text{and} \quad \Delta S^\circ_{subl} = 2.3026(8.566R)$$

$$\Delta H^\circ_{subl} = 129.5 \times 10^3 \text{ J mol}^{-1} \quad \text{and} \quad \Delta S^\circ_{subl} = 164.0 \text{ J K}^{-1}\text{mol}^{-1}$$

Note that "−6762" (and "−4653") in the problem actually are in units of kelvins. The "8.566" (and "5.43") are dimensionless.

(b) A similar analysis for the vaporization of the $NbI_5(l)$ gives

$$\Delta H^\circ_{vap} = 89.08 \times 10^3 \text{ J mol}^{-1} \quad \text{and} \quad \Delta S^\circ_{vap} = 104 \text{ J K}^{-1}\text{mol}^{-1}$$

(c) The normal boiling point of a liquid is defined as the temperature at which the vapor pressure of the liquid equals 1 atm. Therefore:

$$\log 1 = 0 = \frac{-4653}{T_b} + 5.43 \quad \text{hence}: \quad T_b = 857 \text{ K}$$

(d) According to the hint, the vapor pressures of the liquid and solid NbI_5 are equal at the melting point. We simply equate the expressions that give the temperature dependence of these quantities and subscript the temperature as the melting point:

$$\frac{-6762}{T_m} + 8.566 = \frac{-4653}{T_m} + 5.43$$

Solving gives the melting point as 673 K.

11-112 By Le Chatelier's principle, increased pressure favors the state with smaller volume, namely the separate water and salt. The solubility of this salt will decrease as the pressure is increased.

11-114 (a) The chemical amount of CO_2 is

$$n = \frac{PV}{RT} = \frac{(0.972 \text{ atm})(0.20 \text{ L})}{(0.08206 \text{ L atm mol}^{-1}\text{K}^{-1})(373 \text{ K})} = 6.35 \times 10^{-3} \text{ mol}$$

This must have resulted from the same amount of dissolved carbonate in the solution. Thus the concentrations of carbonate and strontium ions were

$$[CO_3^{2-}] = [Sr^{2+}] = \frac{6.35 \times 10^{-3} \text{ mol}}{1.44 \text{ L}} = 4.41 \times 10^{-3} \text{ M}$$

The molar solubility is $4.4 \times 10^{-3} \text{ mol L}^{-1}$.

(b) $K_{sp} = [Sr^{2+}][CO_3^{2-}] = (4.41 \times 10^{-3})^2 = 1.9 \times 10^{-5}$

(c) Strontium and carbonate ions may react with water to form other ionic species. This will change the relationship between K_{sp} and solubility.

Chapter 12

Electrochemistry

12-2 In acidic solution, H_3O^+ and H_2O may be added either as reactants or as products to achieve balance.

(a) $8\,MnO_4^-(aq) + 5\,H_2S(aq) + 14\,H_3O^+(aq) \rightarrow 8\,Mn^{2+}(aq) + 5\,SO_4^{2-}(aq) + 26\,H_2O(l)$

(b) $4\,Zn(s) + NO_3^-(aq) + 10\,H_3O^+(aq) \rightarrow 4\,Zn^{2+}(aq) + NH_4^+(aq) + 13\,H_2O(l)$

(c) $5\,H_2O_2(aq) + 2\,MnO_4^-(aq) + 6\,H_3O^+(aq) \rightarrow 5\,O_2(g) + 2\,Mn^{2+}(aq) + 14\,H_2O(l)$

(d) $2\,Sn(s) + 2\,NO_3^-(aq) + 10\,H_3O^+(aq) \rightarrow 2\,Sn^{4+}(aq) + N_2O(g) + 15\,H_2O(l)$

(e) $3\,UO_2^{2+}(aq) + Te(s) + 4\,H_3O^+(aq) \rightarrow 3\,U^{4+}(aq) + TeO_4^{2-}(g) + 6\,H_2O(l)$

12-4 In basic solution, OH^- and H_2O may take part either as reactants or as products.

(a) $3\,OCl^-(aq) + I^-(aq) \rightarrow IO_3^-(aq) + 3\,Cl^-(aq)$

(b) $2\,SO_3^{2-}(aq) + 2\,Be(s) \rightarrow S_2O_3^{2-}(aq) + Be_2O_3^{2-}(aq)$

(c) $3\,H_2BO_3^-(aq) + 8\,Al(s) + 8\,OH^-(aq) + 7\,H_2O(l) \rightarrow 3\,BH_4^-(aq) + 8\,H_2AlO_3^-(aq)$

(d) $2\,OH^-(aq) + O_2(g) + Sb(s) \rightarrow H_2O_2(aq) + SbO_2^{2-}(aq)$

(e) $2\,Sn(OH)_6^{2-}(aq) + Si(s) \rightarrow 2\,HSnO_2^-(aq) + SiO_3^{2-}(aq) + 5\,H_2O(l)$

12-6 The problem can be completed simply by balancing the equations by the half-reaction method.

(a) oxidation : $12\,OH^-(aq) + 4\,PH_3(g) \rightarrow P_4(s) + 12\,H_2O(l) + 12\,e^-$
reduction : $12\,e^- + 16\,H_2O(l) + 4\,CrO_4^{2-} \rightarrow 4\,Cr(OH)_4^-(aq) + 16\,OH^-(aq)$
The second half-equation in this case can be divided through by 4 to obtain another correct half-equation.

(b) oxidation : $2\,OH^-(aq) + Fe(s) \rightarrow Fe(OH)_2(s) + 2\,e^-$
reduction : $2\,e^- + 2\,H_2O(l) + NiO_2(s) \rightarrow Ni(OH)_2(s) + 2\,OH^-(aq)$

(c) oxidation : $2\,NH_2OH(aq) + 2\,OH^-(aq) \rightarrow N_2(g) + 4\,H_2O(l) + 2\,e^-$
reduction : $CO_2(g) + H_2O(l) + 2\,e^- \rightarrow CO(g) + 2\,OH^-(aq)$

133

12-8 The disproportionation of thiosulfate ion in acidic solution is represented

$$S_2O_3{}^{2-}(aq) + H_3O^+(aq) \rightarrow S(s) + HSO_3^-(aq) + H_2O(l)$$

12-10

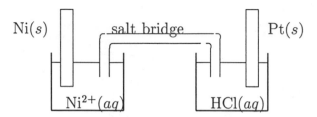

A stream of gaseous hydrogen plays over the platinum electrode. When the two electrodes are then connected by a wire, electrons flow from left to right through the wire. Negative ions migrate to the left in the salt bridge, and positive ions migrate to the right. The overall reaction is:

$$Ni(s) + 2\,HCl(aq) \rightarrow H_2(g) + NiCl_2(aq)$$

12-12 The quantity of electricity is 0.9600 mol of electrons, as shown by dividing the number of coulombs by the Faraday constant. Dissolving 1 mol of Ni to give $Ni^{2+}(aq)$ requires passage of 2 mol of electrons, so the maximum amount of Ni dissolved is 0.4800 mol.

12-14 (a) The balanced equation is $Zn(s) + Cd^{2+}(aq) \rightarrow Zn^{2+}(aq) + Cd(s)$.

(b) The product of the (steady) current in amperes and the time in seconds is the charge in coulombs. It is 1.36×10^4 C. Dividing by the Faraday constant gives the chemical amount of electrons, 0.141 mol.

(c) Every 1 mol of electrons that passes through the cell oxidizes 1/2 mol of zinc. Hence 0.0703 mol of zinc is oxidized. This is 4.60 g of zinc and is the mass lost by the zinc electrode.

(d) Every 1 mol of electrons that passes through the cell reduces 1/2 mol of cadmium(II) ions. This is 0.0703 mol of Cd^{2+}, or 7.91 g of Cd^{2+}. This mass of cadmium metal is gained by the cadmium electrode.

12-16 Multiplying the relative mass of hydrogen liberated by 1 gives close to the known atomic weight of H. Similarly multiplying the given relative mass by 2 for oxygen, by 1 for chlorine and by 2 for tin gives close to the respective relative atomic masses of the elements. The absolute values of the oxidation states of the four elements (hydrogen, oxygen, chlorine, and tin) are therefore in the ratio 1 to 2 to 1 to 2. This follows from Faraday's law that a given amount of charge liberates different substances in proportion to their molar masses (atomic masses) divided by the absolute values of their oxidation numbers.

The hydrogen is liberated at the cathode and must come from the reduction of +1 hydrogen (H^+); other positive oxidation states of hydrogen are nearly unknown. The oxygen is liberated at the *anode*, and so the oxidation state of the oxygen is −2. The oxidation state for chlorine is −1 and that for tin is +2.

12-18 (a) Anode: $C(s) + 2\,O^{2-} \to CO_2(g) + 4e^-$

Cathode: $Al^{3+} + 3e^- \to Al(l)$

Overall: $3\,C(s) + 6\,O^{2-} + 4\,Al^{3+} \to 3\,CO_2(g) + 4\,Al(l)$

(b)

$$\frac{(50,000\ \text{C s}^{-1})(24 \times 60 \times 60\ \text{s})}{96485\ \text{C mol}^{-1}} = 4.48 \times 10^4\ \text{mol}\ e^-$$

$$4.48 \times 10^4\ \text{mol}\ e^- \left(\frac{1\ \text{mol Al}}{3\ \text{mol}\ e^-}\right)(26.98\ \text{g mol}^{-1}) = 4.03 \times 10^5\ \text{g} = 403\ \text{kg Al}$$

12-20 Use the equation:

$$\Delta G = -n\mathcal{F}\Delta\mathcal{E} = w_{elec,max}$$

The $\Delta\mathcal{E}$ is 0.48 V, and $\mathcal{F}$ is 96,485 C mol^{-1}. If 1 mol of zinc were oxidized, then n would equal 2 mol, and $w_{elec,max}$ would be −92.6 kJ. But 1.000 g of zinc is much less than 1.000 mol of zinc; it is only 0.0153 mol. Therefore $n = 0.0306$ mol in the formula, and the maximum work done <u>on</u> the cell is −1.4 kJ. The maximum work done by the cell is the negative of this, or +1.4 kJ per gram of zinc consumed.

12-22 (a) The reduction of Fe^{3+} to Fe^{2+} has $\mathcal{E}^\circ = 0.770$ V, and the reduction of Cd^{2+} to Cd has $\mathcal{E}^\circ = -0.4026$ V. In this galvanic cell with all species in their standard states, the half-reaction $Fe^{3+} + e^- \to Fe^{2+}$ takes place at the cathode, and the oxidation $Cd \to Cd^{2+} + 2\,e^-$ takes place at the anode. Only in this way is $\Delta\mathcal{E}^\circ$ of the cell positive.

(b) $\Delta\mathcal{E}^\circ = \mathcal{E}^\circ$ (cathode) $- \mathcal{E}^\circ$ (anode) $= 0.770 - (-0.4026) = 1.173$ V.

12-24 (a) The increase in the concentration of Cl^- ion means that $Cl_2(g)$ is reduced as the cell operates spontaneously. The half-reaction at the cathode, where reduction occurs, is accordingly $Cl_2 + 2e^- \to 2\,Cl^-$. At the anode, the half-reaction is $Ga \to Ga^{3+} + 3\,e^-$.

(b) The overall voltage is $\Delta\mathcal{E}^\circ$ and equals 1.918 V. It also equals the standard reduction potential of the reduction half-reaction minus the standard reduction potential of the oxidation half-reaction The reduction is the conversion of Cl_2 to Cl^-, which has $\mathcal{E}^\circ$ equal to 1.3583 V (Appendix E). Hence:

$$\Delta\mathcal{E}^\circ = 1.918 = \mathcal{E}^\circ\,(\text{cathode}) - \mathcal{E}^\circ\,(\text{anode}) = 1.3583 - x$$

It follows that x, the reduction potential for the Ga^{3+}/Ga half-cell, is -0.560 V.

12-26 An acidic solution of potassium perchlorate would be an oxidizing agent.

12-28 The standard reduction potentials for ClO_3^-, ClO^-, and Cl_2 in acidic aqueous media are 1.47, 1.63, and 1.3583 V respectively, according to Appendix E. This means that at a given concentration at pH 0, a solution of NaClO is the strongest bleach, despite the fact that the change in oxidation number it experiences is small. At pH 0, the ClO^- ion, a fairly strong base, picks up a hydrogen ion. so the species in the half-equation is HClO. At pH 0 the reduction potential of O_3 to O_2 is 2.07 V (given in Appendix E). Therefore ozone is a stronger bleach than any of the chlorine bleaches.

12-30 (a) H_2O_2 is the strongest oxidizing agent because it is the most easily reduced in this group; it appears near the top left in Appendix E.

(b) Sc is the strongest reducing agent because it is the most easily oxidized in this group; it appears at the bottom right in Appendix E.

(c) Look for a substance on the left side of Appendix E that is between Fe and Cu. The answer is Sn^{2+}.

12-32 (a)
$$\mathcal{E}^\circ = \frac{2(1.25\ \text{V}) - (1)(-0.37\ \text{V})}{1} = 2.87\ \text{V}$$

(b)
$$\Delta\mathcal{E}^\circ = 2.87 - (-0.37) = 3.24\ \text{V} > 0$$

Yes, Tl^{2+} should be disproportionate in aqueous solution.

12-34 (a) ClO^- tends to disproportionate spontaneously at pH 14 to give Cl^- and ClO_2^-. The $\Delta\mathcal{E}^\circ$ of the reaction is $0.90 - 0.59 = 0.31$ V.

(b) According to the reduction potentials given in the problem ClO^- is oxidized to ClO_2^- more readily than Cl^- is oxidized to ClO^- at pH 14; the ClO^- is the stronger reducing agent under these conditions. However, neither of these is a very good reducing agent.

12-36 We calculate the standard potential difference of the cell from the data in Appendix E. At the anode is the Ag/Ag^+ couple; at the cathode is the Cl_2/Cl^- couple. for the overall cell reaction $2\,Ag(s) + Cl_2(g) \rightarrow 2\,Ag^+(aq) + 2\,Cl^-(aq)$. The $\Delta\mathcal{E}^\circ$ is $1.3583 - 0.7996 = 0.5587$ V. This standard voltage must be corrected for the non-standard concentrations in this cell. At 25°C:

$$\Delta\mathcal{E} = \Delta\mathcal{E}^\circ - \frac{0.0592\text{ V}}{n}\log Q = 0.5587\text{ V} - \frac{0.0592\text{ V}}{2}\log\left(\frac{[Ag^+]^2[Cl^-]^2}{P_{Cl_2}}\right)$$

$$\Delta\mathcal{E} = 0.5587\text{ V} - \frac{0.0592\text{ V}}{2}\log\left(\frac{(0.25)^2(0.016)^2}{1.00}\right) = 0.701\text{ V}$$

12-38 We compute the half-cell potential at 25°C using the Nernst equation and the standard reduction potential from Appendix E. The half-reaction is $I_2(s) + 2\,e^- \rightarrow 2\,I^-(aq)$ for which $\mathcal{E}^\circ$ is 0.535 V. Then:

$$\mathcal{E} = \mathcal{E}^\circ - \frac{0.0592\text{ V}}{n_{hc}}\log[I^-]^2$$

$$\mathcal{E} = 0.535\text{ V} - \frac{0.0592\text{ V}}{2}\log[1.5 \times 10^{-6}]^2 = 0.880\text{ V}$$

12-40 We will write the Nernst equation for the redox reaction

$$Cu(s) + Br_2(l) \rightarrow Cu^{2+}(aq) + 2\,Br^-(aq)$$

at 25° C and use it to calculate the unknown concentration. The standard potential difference of the cell comes from the data in Appendix E. It is

$$\Delta\mathcal{E}^\circ = \mathcal{E}\,(\text{cathode}) - \mathcal{E}\,(\text{anode}) = 1.065 - 0.3402 = 0.725\text{ V}$$

The Nernst equation is then:

$$\Delta\mathcal{E} = 0.963\text{ V} = 0.725\text{ V} - \frac{0.0592\text{ V}}{2}\log\left([Cu^{2+}][Br^-]^2\right)$$

Inserting $[Cu^{2+}] = 1.00$ M and solving for $[Br^-]$ gives $[Br^-] = 10^{-4.02} = 9.5 \times 10^{-5}$ M.

12-42 (a) The equations for the reduction (at the cathode) and oxidation (at the anode) are respectively:
$Cr_2O_7^{2-}(aq) + 14\,H_3O^+(aq) + 6\,e^- \rightarrow 2\,Cr^{3+}(aq) + 21\,H_2O(l)$
$6\,I^-(aq) \rightarrow 3\,I_2(s) + 6e^-$
Appendix E provides the standard reduction potentials, and we combine them to

determine the standard voltage of the cell: $\Delta \mathcal{E}^\circ = \mathcal{E}^\circ \text{(cathode)} - \mathcal{E}^\circ \text{(anode)} = 1.33 - 0.535 = 0.795 \approx 0.80$ V.

(b) We write the Nernst equation at 25°C for this cell.

$$\Delta \mathcal{E} = \Delta \mathcal{E}^\circ - \frac{0.0592 \text{ V}}{n} \log Q = \frac{0.0592 \text{ V}}{6} \log \left(\frac{[\text{Cr}^{3+}]^2}{[\text{Cr}_2\text{O}_7^{2-}][\text{I}^-]^6[\text{H}_3\text{O}^+]^{14}} \right)$$

Now, the known voltages and concentrations are inserted:

$$0.87 \text{ V} = 0.795 \text{ V} - \frac{0.0592 \text{ V}}{6} \log \left(\frac{[\text{Cr}^{3+}]^2}{(1.5)\,(0.40)^6\,(1.0)^{14}} \right)$$

Solving gives $[\text{Cr}^{3+}] = 1 \times 10^{-5}$ M.

12-44 The standard potential difference of the reaction is the standard reduction potential for the conversion of mercury(II) to mercury(I) minus the standard reduction potential for the conversion of gold(III) to gold(0): $\Delta \mathcal{E}^\circ = 0.905 - 1.42 = -0.515$ V. The quantity n in the reaction as given in the problem is 6, so at 25°C:

$$\log K = \frac{n}{0.0592 \text{ V}} \Delta \mathcal{E}^\circ = \frac{6}{0.0592 \text{ V}} (-0.515 \text{ V}) = -52.2$$

The equilibrium constant for the reaction as given is $K = 6 \times 10^{-53}$.

The second question really concerns the reverse of the reaction written in the problem. The equilibrium constant of the reverse reaction is the reciprocal of the K just computed:

$$\frac{1}{6 \times 10^{-53}} = \frac{[\text{Hg}^{2+}]^6}{[\text{Au}^{3+}]^2[\text{Hg}_2^{2+}]^3}$$

After the Hg_2^{2+} and Au^{3+} are mixed, but before the reaction starts, their concentrations are both 0.50 M. The equilibrium constant is very large, so the reaction proceeds until the Hg_2^{2+} ion, which is the limiting reactant, is essentially all consumed. Suppose <u>all</u> of the 0.500 M Hg_2^{2+} reacted. The concentration of Hg^{2+} formed is then 1.00 M, and the concentration of excess Au^{3+} is 0.166 M, by the stoichiometry of the reaction. Let the concentration of Hg_2^{2+} actually left at equilibrium equal x. Then $[\text{Au}^{3+}] = 0.166 + 2/3x$ M, and $[\text{Hg}^{2+}] = 1.00 - x/2$ M at equilibrium. We substitute these expressions into the equilibrium law:

$$\frac{1}{6 \times 10^{-53}} = \frac{(1.00 - x/2)^6}{(0.166 + 2/3x)^2(x)^3}$$

Because x is certainly very small compared to 0.166, this equation becomes

$$\frac{1}{6 \times 10^{-53}} \approx \frac{(1.00)^6}{(0.166)^2(x)^3}$$

Solving gives $x = 1 \times 10^{-17}$ M. This is the equilibrium concentration of Hg_2^{2+} ion. The equilibrium concentration of Au^{3+} ion is 0.166 M, and the equilibrium concentration of Hg^{2+} ion is 1.00 M.

12-46 The reaction in the problem is the second of the following half-reactions subtracted from the first

$$Hg^{2+}(aq) + e^- \rightarrow \frac{1}{2}Hg_2^{2+}(aq); \qquad \frac{1}{2}Hg_2^{2+}(aq) + e^- \rightarrow Hg(l)$$

The $\Delta \mathcal{E}^\circ$ for the reaction is accordingly the $\mathcal{E}^\circ$ for the second half-reaction subtracted from the $\mathcal{E}^\circ$ for the first. Taking values from Appendix E, $\Delta\mathcal{E}^\circ = 0.905 - 0.7961 = 0.109$ V. Although the half-equations in Appendix E are written with coefficients that are twice the coefficients in the above, the standard potential difference is the same; it does not depend on the amount of substance that reacts. Inserting this number into the relationship (at 25°C) between the equilibrium constant and the standard potential difference gives:

$$\log_{10} K = \frac{n}{0.0592}\Delta\mathcal{E}^\circ = \frac{1}{0.0592 \text{ V}}(0.109 \text{ V}) = 1.84 \qquad K = 69$$

12-48 The standard potential difference for the reaction is

$$\Delta\mathcal{E}^\circ = \mathcal{E}^\circ(Ag^+/Ag) - \mathcal{E}^\circ(H_3O^+/H_2) = 0.7996 - (0) = 0.7996 \text{ V}$$

. Write the Nernst equation at 25°C

$$\Delta\mathcal{E} = \Delta\mathcal{E}^\circ - \frac{0.0592}{n}\log\frac{[H_3O^+]}{[Ag^+]P_{H_2}^{1/2}}$$

and substitute all of the known values

$$1.030 = 0.7996 - \frac{0.0592}{1}\log\frac{[H_3O^+]}{(1.00)(1.00)}$$

From this equation, $[H_3O^+]$ is 1.28×10^{-4} M. The pH in the buffer solution in the cell is 3.89. Now, compute the K_a of the benzoic acid by substitution in the acid ionization equilibrium expression

$$K_a = \frac{[H_3O^+][C_6H_5COO^-]}{[C_6H_5COOH]} = \frac{(1.28 \times 10^{-4})(0.050)}{(0.10)} = 6.4 \times 10^{-5}$$

12-50 (a) Elemental lead is oxidized, and hydrogen ion is reduced. The standard potential difference is

$$\Delta\mathcal{E}° = \mathcal{E}°(H_3O^+/H_2) - \mathcal{E}°(Pb^{2+}/Pb) = 0 - (-0.1263) = 0.1263 \text{ V}.$$

(b) We write the Nernst equation at 25°C for this cell

$$\Delta\mathcal{E} = \Delta\mathcal{E}° - \frac{0.0592}{n} \log \frac{[Pb^{2+}]P_{H_2}}{[H_3O^+]^2}$$

and substitute all of the known values

$$0.22 = 0.1263 - \frac{0.0592}{2} \log \frac{[Pb^{2+}]1.0}{[1.00]^2}$$

Solving gives the $[Pb^{2+}]$ as 6.8×10^{-4} M.

(c) Presumably the lead concentration in part (b) is in equilibrium with the 0.15 M Cl^- ion. We use the K_{sp} expression

$$K_{sp} = [Pb^{2+}][Cl^-]^2 = (6.8 \times 10^{-4})(0.15)^2 = 1.5 \times 10^{-5}$$

12-52 The overall reaction is

$$Zn(s) + HgO(s) + H_2O(l) \rightarrow Zn(OH)_2(s) + Hg(l)$$

for which

$$\Delta G° = 1(-553.5) + 1(0) - 1(0) - 1(-58.56) - 1(-237.18) = -257.76 \text{ kJ}$$

We compute the $\Delta\mathcal{E}°$ by substitution:

$$\Delta\mathcal{E}° = -\frac{\Delta G°}{n\mathcal{F}} = -\frac{-257.76 \times 10^3 \text{ J}}{(2 \text{ mol})(96485 \text{ C mol}^{-1})} = 1.336 \text{ V}$$

12-54 (a) The reaction is

$$Zn(s) + HgO(s) + H_2O(l) \rightarrow Zn(OH)_2(s) + Hg(l)$$

The equation implies that the reduction of 1 mol of HgO in the battery occurs by passage of 2 mol of electrons through an outside circuit. The 0.50 g of HgO ($\mathcal{M} = 216.59$ g mol^{-1}) is 2.3×10^{-3} mol of HgO. Hence, 4.6×10^{-3} mol of electrons pass the circuit. Multiplying by the Faraday constant converts this to 4.5×10^2 C.

(b) The work performed on the battery (the "system") is

$$w = -Q\Delta\mathcal{E} = -(4.5 \times 10^2 \text{ C})(1.34 \text{ V}) = -6.0 \times 10^2 \text{ J}$$

Thus the battery can perform at most 600 J of work.

12-56 Partial discharge of the lead-acid battery in the vain effort to start the car reduces the concentration of sulfuric acid within the battery. This raises the freezing point of the solution. After the battery re-cools (it generates some internal heat in the starting effort), it can now freeze despite the somewhat warmer ambient temperature.

12-58 The $\Delta G°$ for the oxidation of 1.00 mol of $CO(g)$ to $CO_2(g)$ is -257.21 kJ, regardless of how the reaction is performed. If the reaction is performed with 100% efficiency at standard conditions, then $\Delta G° = w$ and the cell absorbs -257.21 kJ of work, that is, 257.21 kJ of work is obtained.

12-60 The standard cell voltage for the rusting of iron according to the overall equation

$$Fe(s) + 1/2\, O_2(g) + 2\, H_3O^+(aq) \to Fe^{2+}(aq) + 3\, H_2O(l)$$

is $\Delta \mathcal{E}° = \mathcal{E}°_{cathode} - \mathcal{E}°_{anode} = 1.229 - (-0.409) = 1.638$ V. There is a considerable driving force for the rusting of iron at a pressure of $O_2(g)$ of 1 atm and a pH of 0. Making the water even more acidic increases the driving force.

12-62 The successful use of titanium as a sacrificial anode proves that it is more easily oxidized than iron. Assuming that $Ti(s)$ corrodes to $Ti^{3+}(aq)$, the standard reduction potential for Ti^{3+} to $Ti(s)$ is therefore more negative than -0.409 V, which is the standard potential for the reduction of $Fe^{2+}(aq)$ to $Fe(s)$.

12-64 (a) Anode: $3\, H_2O(l) \to \frac{1}{2}\, O_2(g) + 2\, H_3O^+(10^{-7}\ M) + 2\, e^-$

Cathode: $2\, H_3O^+(10^{-7}\ M) + 2\, e^- \to H_2(g) + 2\, H_2O(l)$

(b) $\Delta \mathcal{E}° = -0.414$ V $- 0.815$ V $= -1.229$ V. The decomposition potential is 1.229 V.

(c)

$$\frac{(1.50\ C\ s^{-1})(60 \times 60 \times 50\ s)}{96485\ C\ mol^{-1}} = 2.798\ mol\ e^-$$

$$2.798\ mol\ e^- \left(\frac{1\ mol\ H_2}{2\ mol\ e^-}\right)(2.016\ g\ mol^{-1}) = 2.82\ g\ H_2$$

12-66 The black silver sulfide is reduced $2\, e^- + Ag_2S(s) + H_2O(l) \to 2\, Ag(s) + HS^-(aq) + OH^-(aq)$, and the elemental zinc is oxidized $Zn(s) \to Zn^{2+}(aq) + 2\, e^-$.

12-68 (a) Anode: $Co(s) \to Co^{2+}(aq) + 2\, e^-$
Cathode: $Ag^+(aq) + e^- \to Ag(s)$
Overall: $Co(s) + 2\, Ag^+(aq) \to Co^{2+}(aq) + 2\, Ag(s)$

(b)

$$\frac{0.36 \text{ g}}{58.9332 \text{ g mol}^{-1}} \left(\frac{2 \text{ mol Ag}}{1 \text{ mol Co}}\right) (107.868 \text{ g mol}^{-1}) = 1.32 \text{ g}$$

(c)

$$\frac{0.36 \text{ g}}{58.9332 \text{ g mol}^{-1}} \left(\frac{2 \text{ mol } e^-}{1 \text{ mol Co}}\right) = 0.0122 \text{ mol } e^-$$

$$\frac{(0.0122 \text{ mol})(96485 \text{ C mol}^{-1})}{150 \times 60 \text{ s}} = 0.13 \text{ A}$$

12-70 The key to the problem is writing a balanced redox equation to represent the reaction by which selenate ion is reduced and water is oxidized. The equation is

$$2 \text{ SeO}_4^{2-}(aq) + 2 \text{ H}_2\text{O}(l) \rightarrow 2 \text{ Se}(s) + 3 \text{ O}_2(g) + 4 \text{ OH}^-(aq)$$

It is tempting to focus narrowly on the formula of the selenate ion and reason that 2 mol of $O_2(g)$ must form in the reduction of 1 mol of SeO_4^{2-}. This is wrong because it does not take into account the charge on the selenate ion. The correct ratio is 3/2, as the equation shows.

$$10^{12} \text{ L} \times \left(\frac{0.100 \text{ g SeO}_4^{2-}}{1 \text{ L}}\right) \times \left(\frac{1 \text{ mol SeO}_4^{2-}}{143 \text{ g SeO}_4^{2-}}\right) \times \left(\frac{3 \text{ mol O}_2}{2 \text{ mol SeO}_4^{2-}}\right)$$

$$\times \left(\frac{32.0 \text{ g O}_2}{1 \text{ mol O}_2}\right) = 3.4 \times 10^{10} \text{ g O}_2$$

12-72 The problem can be solved by a series of unit conversions. The maximum allowable concentration of Sn^{2+} is 10 ppm. Each 100 mL sample of rinse solution has a mass of 100 g because it is a dilute solution and consists mostly of water (density $= 1.00$ g mL^{-1}). Then:

$$100 \text{ mL sample} \times \left(\frac{1.00 \text{ g sample}}{1.00 \text{ mL}}\right) \times \left(\frac{10 \times 10^{-6} \text{ g Sn}^{2+}}{1 \text{ g sample}}\right) \times \left(\frac{1 \text{ mol Sn}^{2+}}{118.71 \text{ g Sn}^{2+}}\right)$$

$$\times \left(\frac{2 \text{ mol } e^-}{1 \text{ mol Sn}^{2+}}\right) \times \left(\frac{96485 \text{ C}}{1 \text{ mol } e^-}\right) \times \left(\frac{1 \text{ s}}{25.0 \times 10^{-3} \text{ C}}\right) = 65 \text{ s}$$

12-74 The reaction is

$$\text{TiCl}_4(l) \rightarrow \text{Ti}(s) + 2 \text{ Cl}_2(g)$$

$$\Delta H^\circ = -(-750 \text{ kJ}) = +750 \text{ kJ}$$

$$\Delta S^\circ = 30 + 2(223) - 253 = +223 \text{ J K}^{-1}$$

At 100°C (=373 K),

$$\Delta G = \Delta H^\circ - T\Delta S^\circ = 750,000 \text{ J} - 373 \text{ K}(223 \text{ J K}^{-1}) = +666.8 \text{ kJ mol}^{-1}$$

$$\Delta \mathcal{E}^\circ = -\frac{\Delta G}{n\mathcal{F}} = -\frac{666.8 \times 10^3 \text{ J mol}^{-1}}{(4)(96485 \text{ C mol}^{-1})} = -1.73 \text{ V}$$

The minimum applied voltage is +1.73 V, assuming that $Cl_2(g)$ is produced at 1 atm.

12-76 In the absence of hydrochloric acid, the $\Delta \mathcal{E}^\circ$ for the oxidation of Pt by nitric acid is

$$\Delta \mathcal{E}^\circ = 0.96 - 1.2 \text{ V} = -0.24 \text{ V} < 0$$

The negative sign shows that the reaction

$$3 \text{ Pt} + 2 \text{ NO}_3^- + 8 \text{ H}_3\text{O}^+ \rightarrow 3 \text{ Pt}^{2+} + 2 \text{ NO} + 12 \text{ H}_2\text{O}$$

is not spontaneous.

When hydrochloric acid <u>is</u> present in aqua regia the Pt^{2+} can complex with chloride ion. For the reaction

$$3 \text{ Pt} + 12 \text{ Cl}^- + 2 \text{ NO}_3^- + 8 \text{ H}_3\text{O}^+ \rightarrow 3 \text{ PtCl}_4^{2-} + 2 \text{ NO} + 12 \text{ H}_2\text{O}$$

the cell voltage is

$$\Delta \mathcal{E}^\circ = 0.96 - 0.73 \text{ V} = 0.23 \text{ V} > 0$$

so the reaction can occur spontaneously.

12-78 (a) The $\Delta \mathcal{E}^\circ$ for the proposed reduction of $Cu^{2+}(aq)$ by $I^-(aq$ is $(0.158) - (0.535) = -0.377$ V. The reaction of the two ions in their standard states to give $Cu^+(aq)$ and $I_2(s)$ in their standard states is therefore non-spontaneous.

(b) If we recognize that $CuI(s)$ forms, then the reduction half-reaction in part (a) is replaced by another half-reaction with an $\mathcal{E}^\circ$ of 0.86 V. Now, $\Delta \mathcal{E}^\circ = (0.86) - (0.535) = 0.32$ V. The reaction giving $CuI(s)$ in its standard state from aqueous Cu^{2+} and I^- in their standard states is spontaneous. The formation of insoluble CuI drives an interaction between the ions that would not occur if the product ions formed in their standard states.

12-80 (a) The half-cell potential is calculated using the Nernst equation at 25°C:

$$\mathcal{E} = \mathcal{E}^\circ - \frac{0.0592}{n} \log Q_{\text{hc}} = 1.229 - \frac{0.0592}{4} \log \left(\frac{1}{P_{O_2}[\text{H}_3\text{O}^+]^4} \right)$$

$$\mathcal{E} = 1.229 - \frac{0.0592}{4} \log \left(\frac{1}{(1.0)(1.0 \times 10^{-7})^4} \right) = 0.815 \text{ V}$$

(b) Aeration of a solution puts $O_2(g)$ at a pressure of about 0.2 atm in contact with its contents. The reaction

$$O_2(g) + 4 H_3O^+(aq) + 4 I^-(aq) \rightarrow 6 H_2O(l) + 2 I_2(s)$$

has $\Delta\mathcal{E}° = 0.815 - 0.535 = 0.280$ V at pH 7. This means the oxidation of the iodide ion is spontaneous at pH 7 if there is an 1 atm of oxygen pressure. The reaction tendency is somewhat less at $P_{O_2} = 0.2$ atm, but the reaction is still spontaneous, as can be confirmed with the Nernst equation:

$$\mathcal{E} = 0.280 - \frac{0.0592}{4} \log \frac{1}{P_{O_2}}$$

$$= 1.229 - \frac{0.0592}{4} \log \frac{1}{(0.2)} = 0.280 - 0.01 = 0.27 \text{ V}$$

(c) The standard reduction potentials of $Br_2(l)$ and $Cl_2(g)$ to $Br^-(aq)$ and $Cl^-(aq)$ are larger than that of $I_2(s)$. Both exceed 0.815 V, so the $\Delta\mathcal{E}°$'s for reactions analogous to the aerial oxidation of I^- are negative. Oxygen from the air is not a sufficiently strong oxidizing agent to take $Br^-(aq)$ or $Cl^-(aq)$ to their elemental forms at 1 atm pressure.

(d) Decomposition of solutions of I^- by aeration is favored by increasing the H_3O^+ concentration. This makes the denominator of the Q term in the Nernst equation larger, and $\Delta\mathcal{E}$ is larger.

12-82 (a) Both half-cell reactions involve species not in their standard states. The standard reduction potentials for the two have to be corrected to the actual conditions by means of the Nernst equation. We first compute the reduction potential for the half-reaction at the anode:

$$\mathcal{E}_{Pb^{2+}/Pb} = \mathcal{E}° - \frac{0.0592}{2} \log \frac{1}{[Pb^{2+}]}$$

$$\mathcal{E}_{Pb^{2+}/Pb} = -0.1263 - \frac{0.0592}{2} \log \frac{1}{1.0 \times 10^{-2}} = -0.1855 \text{ V}$$

The half-reaction involving the lead is written as an oxidation in the problem, but a reduction potential must be used in computations. This is why the standard reduction potential appears in the above Nernst equation and why Q_{hc} is set up for a reduction. For the cell:

$$\Delta\mathcal{E} = 0.640 \text{ V} = \mathcal{E} \text{ (cathode)} - \mathcal{E}_{anode} = \mathcal{E} \text{ (cathode)} - (-0.1855)$$

This means that the reduction potential for the VO^{2+}/V^{3+} half-cell at the cathode is 0.4545 V. This is <u>not</u> a standard reduction potential because some of the species in the half-cell are not in their standard states. To compute the standard reduction potential, write the Nernst equation for the VO^{2+}/V^{3+} half-cell:

$$0.4545 = \mathcal{E}^\circ - \frac{0.0592}{n} \log\left(\frac{[V^{3+}]}{[VO^{2+}][H_3O^+]^2}\right) = \mathcal{E}^\circ - \frac{0.0592}{1} \log\left(\frac{(1.0 \times 10^{-5})}{(0.10)(0.10)^2}\right)$$

Solving gives $\mathcal{E}^\circ = 0.336$ V.

(b) The chemical equation in this part is the combination of the two half-reactions of part (a). The standard voltage is

$$\Delta\mathcal{E}^\circ = \mathcal{E}^\circ(\text{cathode}) - \mathcal{E}^\circ(\text{anode}) = 0.336 - (-0.1263) = 0.462 \text{ V}$$

$$\log_{10} K = \frac{n}{0.0592 \text{ V}} \Delta\mathcal{E}^\circ$$

Substitution of $\Delta\mathcal{E}^\circ$ and $n = 2$ gives $K = 4 \times 10^{15}$.

12-84

$$Pb(s) \rightarrow Pb^{2+}(aq) + 2\ e^-$$

$$AgCl(s) + e^- \rightarrow Ag(s) + Cl^-(aq)$$

$$\Delta\mathcal{E}^\circ = 0.222 - (-0.1263) = 0.3483 \text{ V}$$

$$\Delta\mathcal{E} = 0.546 \text{ V} = \Delta\mathcal{E}^\circ - \frac{0.0592}{n} \log Q = 0.3483 - \frac{0.0592}{2} \log[Pb^{2+}][Cl^-]^2$$

Because $[Cl^-] = 1.00$ M,

$$[Pb^{2+}] = 2.1 \times 10^{-7} \text{ M}$$

$$[Pb^{2+}][SO_4^{2-}] = (2.1 \times 10^{-7})(0.0500 \text{ M}) = 1.0 \times 10^{-8} = K_{sp}$$

12-86 The recharging reaction is

$$2\ PbSO_4(s) + 6\ H_2O(l) \rightarrow PbO_2(s) + Pb(s) + 4\ H_3O^+(aq) + 2\ SO_4^{2-}(aq)$$

After this reaction is complete, then the reaction

$$H_2O(l) \rightarrow H_2(g) + 1/2\ O_2(g)$$

begins to occur.

12-88

$$
\begin{aligned}
Fe^{2+} + 2\,e^- &\rightarrow Fe(s) \qquad \mathcal{E}° = -0.409 \text{ V} \\
Sn^{2+} + 2\,e^- &\rightarrow Sn(s) \qquad \mathcal{E}° = -0.1364 \text{ V} \\
Zn^{2+} + 2\,e^- &\rightarrow Zn(s) \qquad \mathcal{E}° = -0.7628 \text{ V}
\end{aligned}
$$

Because Zn^{2+} has a more negative reduction potential, $Zn(s)$ is more easily oxidized thatn $Fe(s)$. It therefore protects the Fe from oxidation. This is not true of tin, however; $Fe(s)$ dissolves, rather then $Sn(s)$, when the film of tin is broken to expose the iron.

12-90

$$
K^+ + e^- \rightarrow K(s) \qquad \mathcal{E}° = -2.925 \text{ V}
$$

This is much more negative than that for

$$
2\,H_3O^+(10^{-7} \text{ M}) + 2\,e^- \rightarrow H_2(g) + 2\,H_2O(l) \qquad \mathcal{E}° = -0.414 \text{ V},
$$

so that $H_2(g)$ will appear at the cathode.

$$
Br_2(l) + 2\,e^- \rightarrow 2\,Br^-(aq) \qquad \mathcal{E}° = 1.065 \text{ V}
$$

$$
\mathcal{E} = \mathcal{E}° - \frac{0.0592}{n}\log[Br^-]^2 = 1.065 - \frac{0.0592}{2}\log(0.05)^2 = 1.142 \text{ V}
$$

This is more positive than that for

$$
2\,H_3O^+(10^{-7} \text{ M}) + \frac{1}{2}O_2(g) + 2\,e^- \rightarrow 3\,H_2O(l) \qquad \mathcal{E}° = 0.815 \text{ V}
$$

so that $O_2(g)$ will appear at the anode.

12-92 The relevant anodic half-reaction is

$$
4\,OH^-(aq) \rightarrow O_2(g) + 2\,H_2O + 4\,e^-
$$

The chemical amount of electrons reacting is

$$
\frac{(0.15 \text{ C s}^{-1})(75 \times 60 \text{ s})}{(96,485 \text{ C mol}^{-1})} = 7.0 \times 10^{-3} \text{ mol } e^-
$$

Because 4 mol e^- accompany 1 mol of $O_2(g)$ generated, the amount of oxygen produced is $\frac{1}{4}$ this, or 1.75×10^{-3} mol O_2. At 25°C the partial pressure of water vapor is 0.03126 atm, so the partial pressure of O_2 is

$$
P_{O_2} = 0.985 - 0.03126 = 0.954 \text{ atm}
$$

and the volume it occupies is

$$V = \frac{n_{O_2}RT}{P_{O_2}} = \frac{(1.75 \times 10^{-3} \text{ mol})(0.08206 \text{ L atm mol}^{-1}\text{K}^{-1})(298 \text{ K})}{0.954 \text{ atm}}$$

$$= 4.5 \times 10^{-2} \text{ L} = 45 \text{ mL}$$

12-94 To make 1.0 kg of Al by recycling cans, the energy cost is

$$\left(\frac{10.7 \text{ kJ mol}^{-1}}{26.98 \text{ g mol}^{-1}}\right) \times (1000 \text{ g kg}^{-1}) = 400 \text{ kJ kg}^{-1}$$

this is only 0.8% of the cost of producing the same kilogram of aluminum from its ore.

12-96 (a) $I_2(aq) + 2 S_2O_3^{2-}(aq) \rightarrow 2 I^-(aq) + S_4O_6^{2-}(aq)$

(b) $(0.05640 \text{ L})(0.100 \text{ mol L}^{-1}) = 5.640 \times 10^{-3} \text{ mol } S_2O_3^{2-}$

$$5.640 \times 10^{-3} \text{ mol } S_2O_3^{2-} \left(\frac{1 \text{ mol } I_2}{2 \text{ mol } S_2O_3^{2-}}\right) = 2.82 \times 10^{-3} \text{ mol } I_2$$

(c) For the reaction

$$I_2(s) + 2 S_2O_3^{2-}(aq) \rightleftharpoons 2 I^-(aq) + S_4O_6^{2-}(aq)$$

$$\Delta\mathcal{E}^\circ = 0.535 - 0.0895 \text{ V} = 0.4455 \text{ V}$$

$$\log_{10} K_1 = \frac{n}{0.0592}\Delta\mathcal{E}^\circ = \frac{2}{0.0592}(0.4455) = 15.05$$

$$K_1 = 1.1 \times 10^{15}$$

For the reaction

$$I_2(aq) \rightleftharpoons I_2(s)$$

$$\Delta G^\circ = 0 - 16.40 \text{ kJ} = -16.40 \text{ kJ}$$

$$K_2 = \exp\left[-\frac{\Delta G^\circ}{RT}\right] = 7.5 \times 10^2$$

The reaction in part (a) is the sum of these two reactions so

$$K = K_1 K_2 = 8.4 \times 10^{17}$$

Chapter 13

Chemical Kinetics

13-2 The rate of the reaction at $t = 100$ s is the slope of a straight line that is tangent to the blue curve in Figure 13.3 at the point that it cuts the vertical line labelled 100 s. If this straight line is sketched, it is seen that it passes through the points $(0.0300, 150)$ and $(0.0125, 0)$. The slope of the line is $(0.0300-0.0125)/(150-0) = 1.17 \times 10^{-4}$ mol $L^{-1}s^{-1}$.

13-4 The expressions are

$$\text{rate} = -\frac{1}{2}\frac{d[H_2CO]}{dt} = -\frac{d[O_2]}{dt} = +\frac{1}{2}\frac{d[CO]}{dt} = +\frac{1}{2}\frac{d[H_2O]}{dt}$$

13-6 (a) The rate equation is

$$\text{rate} = k[SO_2][SO_3]^{-1/2} = k\frac{[SO_2]}{[SO_3]^{1/2}}$$

The reaction is a one-half order reaction overall, so the units of k are $\text{mol}^{1/2}$ $L^{-1/2}$ s^{-1}.

(b) The accelerating effect of the larger concentration of SO_2 is exactly offset by the decelerating effect of the larger concentration of SO_3. The rate does not change.

13-8 (a) Increasing the concentration of Fe^{2+} by a factor of 1.6 between the first and second experiments (while the concentration of Ce^{4+} is constant) increases the rate by a factor of 1.6; the reaction is first-order in Fe^{2+}. Increasing the concentration of Ce^{4+} by a factor of 3.1 between the second and third experiments (while the concentration of Fe^{2+} stays constant) increases the rate by a factor of 3.1; the reaction is first-order in Ce^{4+}. The rate law is rate $= k[Fe^{2+}][Ce^{4+}]$.

148

(b) Substitute one of three sets of data into the rate law. The first set gives:

$$2.0 \times 10^{-7} \text{ mol L}^{-1}\text{s}^{-1} = k \, (1.8 \times 10^{-5} \text{ mol L}^{-1})(1.1 \times 10^{-5} \text{ mol L}^{-1})$$

From this, $k = 1.0 \times 10^3$ L mol^{-1}s^{-1}. The other two sets of data give the same k (to two significant figures).

(c) Substitute the given concentrations and the k from part (b) into the rate law. The initial rate should be 3.4×10^{-7} mol L^{-1}s^{-1}.

13-10 (a) The half-life of a first-order process is 0.6931 (which is ln 2) divided by the first-order rate constant. The answer is 1.03×10^3 s.

(b) A quick answer comes by noting that 0.010 atm is 0.25 times the original pressure of 0.040 atm. The pressure drops to half its original value in the first half-life and to half of that in a second half-life. The answer is two half-lives, or 2.1×10^3 s.

13-12 The partial pressure of the $CH_3CH_2NO_2$ follows the equation

$$\ln P = -kt + \ln P_0$$

because the partial pressure of the gas is directly proportional to its concentration. Substituting gives

$$\ln P = -(1.9 \times 10^{-4} \text{ s}^{-1})(1.08 \times 10^4 \text{ s}) + \ln 0.078 = -4.603$$

Hence $P = 0.010$ atm

13-14 The concentration of the CH_3NC diminishes to 0.71 of its original value in 520 s in a first-order process This means:

$$\ln(0.71) = -kt + \ln(1.00) = k(520 \text{ s}) + 0$$

Solving gives $k = 6.6 \times 10^{-4}$ s^{-1}.

13-16 The disappearance of the HO_2 is second order in the concentration of HO_2. Hence

$$\frac{1}{[HO_2]} - \frac{1}{[HO_2]_0} = 2kt$$

The $[HO_2]_0$ and k are given, so $[HO_2]$ after $t = 1.0$ s is 3.5×10^{-10} mol L^{-1}.

13-18

$$\text{Rate} = -\frac{d[OH^-]}{dt} = k[OH^-][HCN]$$

$$\text{Let } x = [\text{OH}^-] = [\text{HCN}]$$

$$-\frac{dx}{dt} = kx^2$$

$$-\frac{dx}{x^2} = kdt$$

$$\frac{1}{x} - \frac{1}{x_0} = kt$$

$$x_0 = 0.0010 \text{ M}; x = 1.0 \times 10^{-4}$$

$$1.0 \times 10^4 - 1.0 \times 10^3 = kt = 3.7 \times 10^9 \, t$$

$$t = 2.4 \times 10^{-6} \text{ s}$$

13-20 (a) Unimolecular, rate $= k[\text{BrONO}_2]$.

(b) Termolecular, rate $= k[\text{HO}][\text{NO}_2][\text{Ar}]$.

(c) Bimolecular, rate $= k[\text{O}][\text{H}_2\text{S}]$.

13-22 (a) Each of the steps in a mechanism must be an elementary reaction. Therefore, the molecularity of the three steps given here is just the number of species on the left of the arrow: 1, 2, and 3, respectively.

(b) The equation is $\text{NO}_2\text{Cl} + \text{H}_2\text{O} \rightarrow \text{HNO}_3 + \text{HCl}$.

(c) The intermediates are the species in this mechanism that are created in one step of the reaction and consumed in another: NO_2, Cl, OH.

13-24 The equilibrium constant for the mer $\rightleftharpoons$ fac reaction is $K = [\text{fac}]/[\text{mer}]$. At equilibrium the rate of the mer $\rightarrow$ fac reaction equals the rate of the fac $\rightarrow$ mer reaction. Then $2.10 \text{ s}^{-1}[\text{fac}] = 2.33 \text{ s}^{-1}[\text{mer}]$. Solving for the ratio $[\text{fac}]/[\text{mer}]$ gives $K = 1.11$.

13-26 (a) For $2\,\text{A} + 2\,\text{B} \rightarrow \text{E} + \text{G}$ the rate law is: rate $= k_2[\text{D}][\text{B}] = K_1 k_2 [\text{A}]^2 [\text{B}]^2$ according to the mechanism given.

(b) The reaction $\text{A} + \text{B} + \text{D} \rightarrow \text{G}$ has the rate law: rate $= k_3[\text{F}] = k_3 K_2 [\text{C}][\text{D}] = k_3 K_2 K_1 [\text{A}][\text{B}][\text{D}]$ according to the mechanism given.

13-28 The mechanism (a), which starts with a rate-determining collision between $\text{Cl}_2(aq)$ and $\text{H}_2\text{S}(aq)$, is consistent with the observed rate law: rate $= k_1[\text{Cl}_2][\text{H}_2\text{S}]$ According to the second mechanism, the rate would be

$$\text{rate} = k_2[\text{Cl}_2][\text{HS}^-] = k_2 K_1 [\text{Cl}_2]\frac{[\text{H}_2\text{S}]}{[\text{H}^+]}$$

Thus, the reaction would still be first order in both $Cl_2(aq)$ and $H_2S(aq)$, but -1 order in $H^+(aq)$ and therefore only first order overall. This does not fit the stated facts. According to the third mechanism, the reaction would be first order in both $Cl_2(aq)$ and $H_2S(aq)$, but -1 order in $H^+(aq)$ and also -1 order in $Cl^-(aq)$ and therefore zero order overall.

13-30 Both mechanisms (a) and (b) predict the experimentally found rate expression. Mechanism (c) predicts that the reaction would be second order in NO_2 and first order in O_3, and is not acceptable.

13-32

$$\frac{d[C]}{dt} = k_1[A] - k_{-1}[B][C] - k_2[C][D] + k_{-2}[E] = 0$$

$$\frac{d[E]}{dt} = k_2[C][D] - k_{-2}[E] - k_3[E] = 0$$

$$[E] = \frac{k_2[C][D]}{k_{-2} + k_3}$$

Substituting in the first equation gives

$$k_1[A] - k_{-1}[B][C] - k_2[C][D] - \frac{k_{-2}k_2}{k_{-2} + k_3}[C][D] = 0$$

$$[C] = \frac{k_1[A]}{k_{-1}[B] + \left(k_2 + \frac{k_{-2}k_2}{k_{-2}+k_3}\right)[D]}$$

$$[E] = \frac{k_1 k_2 [A][D]}{k_{-1}(k_{-2} + k_3)[B] + k_2(2k_{-2} + k_3)[D]}$$

$$\text{Rate} - k_3[E] = \frac{k_1 k_2 k_3 [A][D]}{k_{-1}(k_{-2} + k_3)[B] + k_2(2k_{-2} + k_3)[D]}$$

This reduces to the result of problem 25b if $k_3 << k_{-2}$ and $k_2[D] << k_{-1}[B]$.

13-34 According to the steady-state approximation,

$$\frac{d[SO_2OOH^-]}{dt} = k_1[HSO_3^-][H_2O_2] - k_{-1}[SO_2OOH^-] - k_2[SO_2OOH^-][H_3O^+] = 0$$

Solving for the concentration of SO_2OOH^- gives

$$[SO_2OOH^-] = \frac{k_1[HSO_3^-][H_2O_2]}{k_{-1} + k_2[H_3O^+]}$$

so that the reaction rate is

$$\text{rate} = k_2[SO_2OOH^-][H_3O^+] = \frac{k_1 k_2 [HSO_3^-][H_2O_2][H_3O^+]}{k_{-1} + k_2[H_3O^+]}$$

13-36 According to the Arrhenius equation, the logarithm of the rate constant is a linear function of the reciprocal of the absolute temperature with a slope of $-E_a/R$ and an intercept of $\ln A$.

(a) Most students will use the Arrhenius equation with various pairs of the six data given and compute varied answers, because there is some experimental scatter to the points. The best way to use all of the data is to perform a linear least-square fit of $\ln k$ versus $1/T$. The slope of the line is -4360 K, and the intercept is 23.962. This means that $E_a = -(-4360 \text{ K})(8.315 \text{ J mol}^{-1} \text{ K}^{-1}) = 36,200 \text{ J mol}^{-1}$.

(b) The constant A is the exponential of the intercept from part (a). It always has the same units as k: $A = 2.6 \times 10^{10} \text{ s}^{-1}$.

13-38 (a)

$$-\ln \frac{x}{x_0} = kt$$

$$-\ln \frac{0.010 \text{ atm}}{0.10 \text{ atm}} = (5.1 \times 10^6 \text{s}^{-1})t$$

$$t = 4.5 \times 10^{-7}\text{s}$$

(b)

$$\ln \frac{k_{300}}{5.1 \times 10^6} = \frac{-54.0 \times 10^3 \text{ J mol}^{-1}}{8.315 \text{ J mol}^{-1} \text{ K}^{-1}} \left[\frac{1}{573.15 \text{ K}} - \frac{1}{303.15 \text{ K}} \right]$$

$$k_{300} = 1.2 \times 10^{11} \text{ s}^{-1}$$

$$-\ln \left(\frac{1}{10} \right) = 1.2 \times 10^{11} \text{ s}^{-1}t$$

$$t = 1.9 \times 10^{-11} \text{ s}$$

13-40 (a) The Arrhenius equation gives $\ln A$ in terms of k, T, and E_a, which are stated explicitly in the problem:

$$\ln A = \ln k + \frac{E_a}{RT} = \ln(6.1 \times 10^{-4}) + \frac{272,000 \text{ J mol}^{-1}}{(8.315 \text{ J mol}^{-1} \text{ K}^{-1})(773 \text{ K})} = 34.9$$

Hence $A = 1.5 \times 10^{15} \text{ s}^{-1}$. Note that the units of A are always the same as the units of k. Also, the units in the activation energy term cancel out completely.

(b) Writing the Arrhenius equation at the two temperatures and taking the ratio gives

$$\ln \frac{k_{773}}{k_{298}} = -\frac{272,000 \text{ J}}{8.315 \text{ J mol}^{-1} \text{ K}^{-1}} \left(\frac{1}{773 \text{ K}} - \frac{1}{298 \text{ K}} \right)$$

The ratio k_{773}/k_{298} equals 2×10^{29}, and k_{773} is given. Therefore $k_{298} = 3.1 \times 10^{-33} \text{ s}^{-1}$. The reaction is very slow at room conditions.

13-42 The ΔE for the HOCl $\rightarrow$ HClO conversion should equal the activation energy for the forward reaction minus the activation energy for the reverse reaction, or $311 - 31 = 280$ kJ mol^{-1}.

13-44 The preexponential factor is $2d^2 N_0 \sqrt{\frac{\pi RT}{M}} P$. Substituting $d = 2.6 \times 10^{-10}$m, $N_0 = 6.022 \times 10^{23}$ mol^{-1}, $R = 8.315$ J mol^{-1} K^{-1}, $T = 500$ K, $M = 46.01 \times 10^{-3}$ kg mol^{-1}, and $P = 5.0 \times 10^{-2}$ gives a result of

$$2.2 \times 10^6 \text{ m}^3 \text{ mol}^{-1} \text{ s}^{-1} = 2.2 \times 10^9 \text{ L mol}^{-1} \text{ s}^{-1}.$$

13-46 (a) The maximum rate of reaction occurs at very high substrate concentration. In this limit, the Michaelis-Menten rate becomes

$$\text{rate} = k_2[E]_0 = (6 \times 10^5 \text{ s}^{-1})(5 \times 10^{-6} \text{ mol L}^{-1}) = 3 \text{ mol L}^{-1} \text{ s}^{-1}$$

(b) We can write

$$\text{rate} = \frac{(\text{max rate})[S]}{[S] + K_m}$$

$$\frac{\text{rate}}{\text{max rate}} = \frac{[S]/K_m}{1 + [S]/K_m} = 0.30 \ (30\%)$$

Solving gives $[S]/K_m = 0.43$

$$[S] = 3.4 \times 10^{-5} \text{ M}$$

13-48 This problem treats an instance of competing reactions reported in the literature (Inorg. Chem. **26** 948 (1987)).

(a) The thiosulfate ion and hydrogen peroxide interact in two different ways. Let x be the initial rate of disappearance of $S_2O_3^{2-}$ by the first reaction and let $2y$ be the initial rate of disappearance of $S_2O_3^{2-}$ by the second reaction. Then $4x$ is the rate of disappearance of H_2O_2 by the first reaction and y is the rate of disappearance of H_2O_2 by the second reaction. These relationships follow from the stoichiometry. It is now possible to write:

$$x + 2y = 7.9 \times 10^{-7} \text{ mol L}^{-1}\text{s}^{-1} \quad \text{and} \quad 4x + y = 8.8 \times 10^{-7} \text{ mol L}^{-1}\text{s}^{-1}$$

because the total rates of disappearance of the reactants are the sums of the rates via the different routes. Solving gives

$$x = 1.386 \times 10^{-7} \text{ mol L}^{-1}\text{s}^{-1} \quad \text{and} \quad y = 3.257 \times 10^{-7} \text{ mol L}^{-1}\text{s}^{-1}$$

The percentage of $S_2O_3^{2-}$ reacting at the first moment according to the first equation is the rate x divided by total rate of disappearance of $S_2O_3^{2-}$ times 100%. It is 18%.

(b) The first reaction generates H_3O^+ at the initial rate $2x$, but the second reaction consumes H_3O^+ at the initial rate $2y$. The net rate of consumption of H_3O^+ is

$$2y - 2x = 3.742 \times 10^{-7} \text{ mol L}^{-1}\text{s}^{-1}$$

This rate is also the initial rate at which replacement H_3O^+ has to be added to keep the pH at 7.0. The volume of the solution to which it is added is 2.00 L, so 7.484×10^{-7} mol s^{-1} of H_3O^+ has to be added. In 60 s, this amounts to 4.49×10^{-5} mol. This much H_3O^+ is contained in 0.45 mL of 0.100 M H_3O^+.

13-50 (a) The initial partial pressure (the pressure at time 0) of DTBP is 0.2362 atm. If it has decreased by x atm at time t, to a value of $0.2362 - x$ atm, then the total pressure of the gas mixture at time t is $0.2362 + 2x$. Then we have:

$$P_{\text{tot}} = 0.2362 + 2x \text{ atm} \quad \text{so} \quad x = \frac{1}{2}P_{\text{tot}} - 0.1181 \text{ atm}$$

The expression for the partial pressure of DTBP is:

$$P_{\text{DTBP}} = 0.2362 - x = 0.2362 - \left(\frac{1}{2}P_{\text{tot}} - 0.1181\right) = 0.3543 - \frac{1}{2}P_{\text{tot}} \text{ atm}$$

The following table shows the computed values for the pressure of DTBP based on the total pressure at the various times:

time (s)	P_{DTBP} (atm)	time (s)	P_{DTBP} (atm)
0	0.2362	26	0.1882
2	0.2310	30	0.18185
6	0.22365	34	0.1758
10	0.2158	38	0.16995
14	0.20875	40	0.16685
18	0.20175	42	0.16425
20	0.1982	46	0.15885
22	0.1949		

(b) The partial pressure of DTBP and its concentration are in direct proportion. If the reaction is first order in DTBP, then, over any set period of time, the natural logarithm of the DTBP pressure will always change by the same amount. If the reaction is second order in DTBP, then, instead of $\ln P_{\text{DTBP}}$, $1/P_{\text{DTBP}}$ will exhibit a constant change over equal time intervals. The interval of 4 minutes is common in the table. From 2 to 6 minutes $\Delta(1/P_{\text{DTBP}})$ is 0.1423; from 42 to

46 minutes, it is 0.2070. This is a substantial alteration. For the same pair of intervals $\Delta(\ln P_{\text{DTBP}}) = -0.0323$ and -0.0334. This is very little change. The reaction is first order in DTBP. The same result is obtained by plotting the full set of data as in text Figure 13.8.

13-52 The forward rate of the reaction and the reverse rate of the reaction are

$$\text{rate}_f = k_f P_{\text{CO}_2} P_{\text{NH}_3}^2 \quad \text{and} \quad \text{rate}_r = k_r$$

The net rate is the forward rate minus the reverse rate. At equilibrium the net rate is zero. Hence

$$k_f P_{\text{CO}_2} P_{\text{NH}_3}^2 = k_r$$

and

$$\frac{k_f}{k_r} = \frac{1}{P_{\text{CO}_2} P_{\text{NH}_3}^2}$$

But the expression on the right is the equilibrium constant, so $K = k_f/k_r$. Substitution of the two k's gives $K = 1.49 \times 10^6$.

13-54 (a) At equilibrium, $\frac{[B]}{[A]} = \frac{k_1}{k_{-1}}$ and $\frac{[C]}{[A]} = \frac{k_2}{k_{-2}}$. Dividing the second expression by the first gives

$$\frac{[C]}{[B]} = \frac{k_2\, k_{-1}}{k_{-2} k_1} = \frac{(1 \times 10^9)(1 \times 10^2)}{(1 \times 10^4)(1 \times 10^8)} = 0.1 = K$$

The ratio of [B] to [C] is the inverse of this, or 10.

(b) If k_{-1} and k_{-2} can be ignored, then B forms at a rate k_1/k_2 times that of C. In this case,

$$\frac{[B]}{[C]} = \frac{k_1}{k_2} = \frac{1 \times 10^8}{1 \times 10^9} = 0.1$$

13-56 A plot of k_{eff}^{-1} against P^{-1} is linear. Using least squares we find the intercept to be 3.316, which gives

$$\frac{k_1 k_2}{k_{-1}} = 0.302 \text{ s}^{-1}$$

The slope is 3.321 atm s. Converting from pressure to concentration using $P/RT = [M]$ gives

$$k_1 = 7.4 \text{ L mol}^{-1} \text{ s}^{-1}$$

13-58 (a)

$$\frac{d[H]}{dt} = k_2[\text{Br}][\text{H}_2] - k_4[\text{HBr}][\text{H}] - k_3[\text{H}][\text{Br}_2]$$

(b)

$$\frac{d[\text{Br}]}{dt} = 2\,k_1[\text{Br}_2][\text{M}] - 2\,k_{-1}[\text{Br}]^2[\text{M}] - k_2[\text{Br}][\text{H}_2] + k_4[\text{HBr}][\text{H}] + k_3[\text{H}][\text{Br}_2]$$

Note that $2\,k_1$ and $2\,k_{-1}$ appear because two Br atoms are formed from each Br_2 molecule.

(c) Setting the two derivatives to zero and adding gives

$$2\,k_1[\text{Br}_2][\text{M}] - 2\,k_{-1}[\text{Br}]^2[\text{M}] = 0$$

$$[\text{Br}] = \sqrt{\frac{k_1[\text{Br}_2][\text{M}]}{k_{-1}[\text{M}]}} = \sqrt{\frac{k_1}{k_{-1}}}[\text{Br}_2]^{1/2}$$

Setting $\dfrac{d[\text{H}]}{dt} = 0$ and solving for [H] gives

$$[\text{H}] = \frac{k_2[\text{Br}][\text{H}_2]}{k_4[\text{HBr}] + k_3[\text{Br}_2]} = \frac{k_2(k_1/k_{-1})^{1/2}[\text{H}_2][\text{Br}_2]^{1/2}}{k_4[\text{HBr}] + k_3[\text{Br}_2]}$$

(d)

$$\frac{d[\text{HBr}]}{dt} = k_2[\text{Br}][\text{H}_2] + k_3[\text{H}][\text{Br}_2] - k_4[\text{HBr}][\text{H}]$$

$$= k_2(k_1/k_{-1})^{1/2}[\text{H}_2][\text{Br}_2]^{1/2} + \frac{k_3[\text{Br}_2] - k_4[\text{HBr}]}{k_3[\text{Br}_2] + k_4[\text{HBr}]}k_2(k_1/k_{-1})^{1/2}[\text{H}_2][\text{Br}_2]^{1/2}$$

$$= \frac{2\,k_3k_2(k_1/k_{-1})^{1/2}[\text{H}_2][\text{Br}_2]^{3/2}}{k_3[\text{Br}_2] + k_4[\text{HBr}]}$$

13-60 The logarithm of the rate constant k is a linear function of the reciprocal of the absolute temperature. The slope of this line on an Arrhenius plot is $-E_a/R$. Converting the Celsius temperatures to Kelvin temperatures and performing a least-squares fit (the best way to use all four data points, see solution to problem 13-36) gives slope $= -7606.6$ K, and therefore $E_a = 63.2$ kJ mol^{-1}.

13-62 (a) Let the ratio of the rates at 308 K and 298 K be x. Then, in the Arrhenius equation:

$$\ln x = \frac{-53,000 \text{ J mol}^{-1}}{8.315 \text{ J mol}^{-1} \text{ K}^{-1}}\left(\frac{1}{308 \text{ K}} - \frac{1}{298 \text{ K}}\right)$$

Solving gives $x = 2.00$.

(b) A similar substitution in the Arrhenius equation gives:

$$\ln x = \frac{-53,000 \text{ J mol}^{-1}}{8.315 \text{ J mol}^{-1} \text{ K}^{-1}}\left(\frac{1}{408 \text{ K}} - \frac{1}{398 \text{ K}}\right)$$

Solving gives $x = 1.48$, which is 1.5.

13-64 (a) The reactants are A_2, B, and CD, and the products are AC and BD. The balanced equation is $A_2 + 2\,B + 2\,CD \rightarrow 2\,AC + 2\,BD$. Note that the second step occurs twice as often as the first.

(b) Rate $= k_3[AB][CD] = k_3K_2[A][B][CD] = k_3K_2K_1^{1/2}[A_2]^{1/2}[B][CD]$.

(c) Because the first two steps are endothermic, K_1 and K_2 must increase with temperature. This is also true for any elementary reaction rate constant such as k_3. The result is to increase the overall reaction rate constant.

13-66

$$\frac{d[Ag^{2+}]}{dt} = k_1[Ag^+][Ce^{4+}] - k_{-1}[Ag^{2+}][Ce^{3+}] - k_2[Tl^+][Ag^{2+}] = 0$$

$$[Ag^{2+}] = \frac{k_1[Ag^+][Ce^{4+}]}{k_{-1}[Ce^{3+}] + k_2[Tl^+]}$$

$$\text{Rate} = k_2[Tl^+][Ag^{2+}] = \frac{k_1k_2[Ag^+][Ce^{4+}][Tl^+]}{k_{-1}[Ce^{3+}] + k_2[Tl^+]}$$

13-68 The maximum rate is $k_2[E]_0$, so

$$1 \times 10^{-6} \text{ mol L}^{-1} \text{ s}^{-1} = k_2(2 \times 10^{-6} \text{ mol L}^{-1})$$

$$k_2 = 0.5$$

$$\frac{\text{rate}}{\text{max rate}} = \frac{1}{2} = \frac{[S]}{[S] + K_m}$$

Inserting $[S] = 6 \times 10^{-6}$ M and solving for K_m gives

$$K_m = 6 \times 10^{-6} \text{ mol L}^{-1}$$

13-70 Acid-base equilibria are rapidly established, so we can write

$$[H_3O^+] = K_a\frac{[HCN]}{[CN^-]} = (6.17 \times 10^{-10}) \times \frac{0.095 \text{ M}}{0.17 \text{ M}} = 3.45 \times 10^{-10} \text{ M}$$

$$[OH^-] = \frac{K_w}{[H_3O^+]} = 2.9 \times 10^{-5} \text{ M}$$

The initial rate is

$$\text{rate} = k[ClO_2]^2[OH^-]$$

$$= (230 \text{ L}^2 \text{ mol}^{-2}\text{s}^{-1})(0.020 \text{ mol L}^{-1})^2(2.9 \times 10^{-5} \text{ mol L}^{-1})$$

$$= 2.7 \times 10^{-6} \text{ mol L}^{-1}\text{s}^{-1}$$

13-72 The rms velocity is

$$u_{\text{rms}} = \sqrt{\frac{3RT}{\mathcal{M}}}$$

which is equal to 466 m s^{-1}. For a molecule with kinetic energy E_a/N_0,

$$\frac{1}{2}mv^2 = \frac{E_a}{N_0}$$

$$v^2 = 2\frac{E_a}{mN_0} = 2\frac{E_a}{\mathcal{M}}$$

Solving for v gives

$$v = 2197 \text{ m s}^{-1} \approx 2200 \text{ m s}^{-1}$$

which is larger by a factor of 4.7.

Chapter 14

Nuclear Chemistry

14-2 (a) $^{4}_{2}\text{He} + ^{253}_{99}\text{Es} \rightarrow ^{255}_{101}\text{Md} + 2\,^{1}_{0}n$

(b) $^{249}_{98}\text{Cf} + ^{10}_{5}\text{B} \rightarrow ^{257}_{103}\text{Lr} + 2\,^{1}_{0}n$

(c) $^{238}_{92}\text{U} + ^{12}_{6}\text{C} \rightarrow ^{244}_{98}\text{Cf} + 6\,^{1}_{0}n$

14-4 The mass of each atom minus the sum of the masses of its constituent protons, neutrons, and electrons is the mass change as the atom forms from its constituent particles. This difference in mass is converted from units of mass to units of energy using the knowledge that 1 amu is equivalent to 931.494 MeV, and 1 MeV equals 1.602177×10^{-13} J. Binding energies are the negatives of the ΔE's, and the binding energies per nucleon are the binding energies per atom divided by A, the number of protons plus neutrons.

(a) For formation of <u>atoms</u> of $^{10}_{4}\text{Be}$: the Δm is -0.0697546 u;
$E_b = 64.976$ MeV per atom $= 6.26923 \times 10^9$ kJ mol^{-1};
binding energy per nucleon $= 64.976/10 = 6.4976$ MeV per nucleon.

(b) For formation of atoms of $^{35}_{17}\text{Cl}$: Δm is -0.3201414 u;
$E_b = 298.210$ MeV per atom $= 28.7729 \times 10^9$ kJ mol^{-1};
binding energy per nucleon $= 298.210/35 = 8.52028$ MeV per nucleon.

(c) For formation of atoms of $^{49}_{22}\text{Ti}$: Δm is -0.4582324 u;
$E_b = 426.841$ MeV per atom $= 41.1839 \times 10^9$ kJ mol^{-1};
binding energy per nucleon $= 426.841/49 = 8.71104$ MeV per nucleon.

14-6 We compare the mass of two $^{16}_{8}\text{O}$ atoms to the mass of one $^{32}_{16}\text{S}$ atom. The S atom has a smaller mass, so it is more stable. Subtracting the mass of two O's from the mass of one S gives $\Delta m = -0.017757$ u.

14-8 The nuclear reaction is $^{10}_{4}\text{Be} \rightarrow ^{10}_{5}\text{B} + ^{0}_{-1}e^- + \bar{\nu}$. The kinetic energy of the $^{0}_{-1}e^-$ is a maximum when the antineutrino has zero kinetic energy. In this case, essentially all of the ΔE of the reaction is carried away by the $_{-1}^0e^-$. The ΔE is $c^2\Delta m$, and $\Delta m = -0.000597$ u, so $\Delta E = -0.556$ MeV. The maximum energy of the $_{-1}^0e^-$ is $+0.556$ MeV. Remember that the mass of the $_{-1}^0e^-$ is *not* added in calculating Δm

14-10 (a) $_{70}^{155}\text{Yb} \rightarrow {}_{68}^{151}\text{Er} + {}_2^4\text{He}$ (b) $_{14}^{26}\text{Si} \rightarrow {}_{13}^{26}\text{Al} + {}_1^0e^+ + \nu$
(c) $_{30}^{65}\text{Zn} + {}_{-1}^0e^- \rightarrow {}_{29}^{65}\text{Cu} + \nu$ (d) $_{41}^{100}\text{Nb} \rightarrow {}_{42}^{100}\text{Mo} + {}_{-1}^0e^- + \bar{\nu}$

14-12 $_6^{11}\text{C} \rightarrow {}_5^{11}\text{B} + {}_1^0e^+$

$_6^{14}\text{C} \rightarrow {}_7^{14}\text{N} + {}_{-1}^0e^-$

Alpha emission would yield $_4^{10}\text{Be}$ via

$$_6^{14}\text{C} \rightarrow {}_4^{10}\text{Be} + {}_2^4\text{He}$$

$$\Delta m = 0.0129 \text{ u} > 0$$

This is not a spontaneous process.

14-14 (a) The daughter nuclide is just the parent minus one alpha particle ($_2^4\text{He}$). The nuclear reaction for this loss is $_{84}^{210}\text{Po} \rightarrow {}_{82}^{206}\text{Pb} + {}_2^4\text{He}$.

(b) The Δm in this reaction is the mass of the daughter atom plus the mass of a helium-4 atom minus the mass of the parent atom; it is -0.0057967 u. The energy released is $-\Delta E = -c^2\Delta m = 5.40$ MeV

(c) The kinetic energy of the alpha particle is approximately equal to the energy released, or 8.65×10^{-13} J.

14-16 $_{83}^{209}\text{Bi} + {}_0^1n \rightarrow {}_{83}^{210}\text{Bi}$ $_{83}^{210}\text{Bi} \rightarrow {}_{84}^{210}\text{Po} + {}_{-1}^0e^- + \bar{\nu}$

14-18 (a) $_{83}^{209}\text{Bi} + {}_{28}^{64}\text{Ni} \rightarrow {}_{111}^{272}\text{Uuu} + {}_0^1n$ (b) $_{111}^{272}\text{Uuu} \rightarrow {}_{109}^{268}\text{Mt} + {}_2^4\text{He}$

14-20 We compute the half-life of ^{238}U in minutes:

$$t_{1/2} = 4.47 \times 10^9 \text{ yr} \times \left(\frac{365 \text{ day}}{1 \text{ yr}}\right) \times \left(\frac{1440 \text{ min}}{1 \text{ day}}\right) = 2.35 \times 10^{15} \text{ min}$$

The activity equals the decay constant of the radioactive nuclide times the number of those nuclides in the sample ($A = kN$). The decay constant equals the half-life divided into the natural logarithm of 2.

$$A = kN = \left(\frac{\ln 2}{t_{1/2}}\right)\left(\frac{0.0010 \text{ g U}}{238 \text{ g mol}^{-1}}\right) \times \left(\frac{6.022 \times 10^{23} \text{ atom U}}{1 \text{ mol U}}\right)$$

$$= \left(\frac{0.6931}{2.35 \times 10^{15} \text{ min}}\right) \times (2.53 \times 10^{18} \text{ atom U}) = 7.46 \times 10^2 \frac{\text{atoms U}}{\text{min}}$$

14-22 (a) The decay constant equals ln 2 divided by the half-life. In this case, it is $(0.6931/87.1 \text{ day}) = 0.007958 \text{ day}^{-1}$. Two of the three quantities in the equation $A = kN$ are now known. The third is:

$$N = \frac{A}{k} = \left(\frac{3.70 \times 10^2 \text{ s}^{-1}}{0.00796 \text{ day}^{-1}}\right) \times \left(\frac{86,400 \text{ day}^{-1}}{1 \text{ s}^{-1}}\right) = 4.017 \times 10^9$$

This number of atoms of ^{35}S is converted first to moles and then to grams

$$4.017 \times 10^9 \text{ atoms} \times \left(\frac{1 \text{ mol S}}{6.022 \times 10^{23} \text{ atoms}}\right) \times \left(\frac{35 \text{ g } ^{35}\text{S}}{1 \text{ mol } ^{35}\text{S}}\right) = 2.3 \times 10^{-13} \text{ g } ^{35}\text{S}$$

(b) We know the decay constant of the radioactive isotope, we know how much of it we start with, and we know that the decay goes on for 365 days.

$$N = N_i e^{-kt} = 4.017 \times 10^9 \exp\left[-(0.007958 \text{ day}^{-1})(365 \text{ day})\right] = 2.20 \times 10^8 \text{ atoms}$$

This number of atoms of ^{35}S is a very small mass, only 1.3×10^{-14} g.

14-24 The 1.0 μg of ^{99m}Tc contains 6.08×10^{15} atoms, a result obtained by dividing the 1.0 μg by the molar mass of 99 g mol^{-1} and multiplying by Avogadro's number. The 6.0 hr half-life is 2.16×10^4 s. Dividing this into ln 2 gives the decay constant $k = 3.21 \times 10^{-5}$ s^{-1}. The product of the number of atoms and the decay constant is the activity of the sample. It is 2.0×10^{11} disintegrations s^{-1}, or 5.4 Ci.

14-26

$$A = A_0 \exp(-kt)$$

$$0.0375 \text{ Bq g}^{-1} = (0.255 \text{ Bq g}^{-1}) \exp(-kt)$$

$$kt = 1.917$$

$$t = \frac{1.917}{k} = \frac{1.917}{\ln 2} t_{\frac{1}{2}} = 15,800 \text{ years}$$

14-28

$$\text{moles He} = \frac{(4.9 \times 10^{-3} \text{ cm}^3)(10^{-3} \text{ L cm}^{-3})(1 \text{ atm})}{(0.08206 \text{ L atm mol}^{-1} \text{ K}^{-1})(273.15 \text{ K})} = 2.2 \times 10^{-7} \text{ mol}$$

$$\text{moles } ^{232}\text{Th decaying} = \frac{\text{moles He}}{6} = 3.64 \times 10^{-8} \text{ mol}$$

$$\text{moles } ^{232}\text{Th remaining} = \frac{7.4 \times 10^{-3} \text{ g}}{232 \text{ g mol}^{-1}} = 3.19 \times 10^{-5} \text{ mol}$$

$$\frac{3.19 \times 10^{-5}}{3.19 \times 10^{-5} + 3.64 \times 10^{-8}} = e^{-kt}$$

$$kt = 1.14 \times 10^{-3}$$

$$t = \frac{1.14 \times 10^{-3}}{k} = \frac{1.14 \times 10^{-3}}{\ln 2}(1.39 \times 10^{10} \text{ yr}) = 2.3 \times 10^{7} \text{yr}$$

The sediment is about 23 million years old.

14-30 From 14-29 we found that the ratio of the numbers of ^{238}U and ^{235}U isotopes was unity at $t = 6.02 \times 10^{9}$yr ago, when a supernova created the solar system. At $t = (6.02 - 4.50) \times 10^{9}$ yr after the birth of the solar system:

$$\ln\left(\frac{N_{235_U}}{N_0}\right) = -k_{235} \times 1.52 \times 10^{9} \text{ yr}$$

and

$$\ln\left(\frac{N_{238_U}}{N_0}\right) = -k_{238} \times 1.52 \times 10^{9} \text{ yr}$$

$$\ln\left(\frac{N_{238_U}}{N_{235_U}}\right) = -(k_{238} - k_{235}) \times 1.52 \times 10^{9} \text{ yr}$$

$$k_{238} = \frac{0.6931}{4.47 \times 10^{9} \text{ yr}} \quad \text{and} \quad k_{235} = \frac{0.6931}{7.04 \times 10^{8} \text{ yr}}$$

$$\ln\left(\frac{N_{238_U}}{N_{235_U}}\right) = 0.6931\left(\frac{1}{7.04 \times 10^{8}} - \frac{1}{4.47 \times 10^{9}}\right) \times 1.52 \times 10^{9} \text{ yr}$$

$$\frac{N_{238_U}}{N_{235_U}} = 3.5 \quad \text{when the earth was formed.}$$

14-32 Obviously, the decay must be by positron emission:
$$^{13}_{7}\text{N} \rightarrow {}^{13}_{6}\text{C} + {}^{0}_{1}e^{+} + \nu \quad \text{and} \quad {}^{18}_{9}\text{F} \rightarrow {}^{18}_{8}\text{O} + {}^{0}_{1}e^{+} + \nu.$$
The positrons are annihilated in collisions with electrons in ordinary matter.

14-34 The ^{226}Ra emits more energetic particles and at a greater rate. The emitted particles are moreover absorbed more efficiently. The ^{226}Ra is therefore far more dangerous than the ^{14}C.

14-36 (a) The number of atoms of ^{239}Pu ingested is

$$N = \left(\frac{5.0 \times 10^{-6} \text{ g}}{239 \text{ g mol}^{-1}}\right)\left(6.022 \times 10^{23} \text{ atoms mol}^{-1}\right) = 1.26 \times 10^{16} \text{ atoms}$$

The activity is

$$A = kN = \frac{N \ln 2}{t_{1/2}} = \frac{(1.26 \times 10^{16})(0.6931)}{(2.411 \times 10^4 \text{ yr})(365 \times 24 \times 60 \times 60 \text{ s yr}^{-1})}$$

$$= 1.15 \times 10^4 \text{ s}^{-1} = 1.15 \times 10^4 \text{ Bq}$$

(b) The energy dissipated per year is

$$(1.15 \times 10^4 \text{ s}^{-1})(60 \times 60 \times 24 \times 365 \text{ s yr}^{-1})(5.24 \text{ MeV}) = 1.9 \times 10^{12} \text{ MeV yr}^{-1}$$

Convert this to centijoules (1 cJ $= 10^{-2}$ J):

$$(1.9 \times 10^{12} \text{ MeV yr}^{-1})(1.602 \times 10^{-13} \text{ J MeV}^{-1})(10^2 \text{ cJ J}^{-1}) = 30.4 \text{ cJ yr}^{-1}$$

Each kilogram of tissue receives 1/60 of this, because the worker weighs 60 kg. The dose is

$$\frac{30.4}{60} = 0.51 \text{ cJ kg}^{-1} \text{ yr}^{-1} = 510 \text{ mrad yr}^{-1}$$

(c) Convert this dose to mrem by using the statement on p. 464 that a dose of 1 rad of alpha radiation is equivalent to 10 rems. The result is 5,100 mrem yr^{-1}. This is about fifty times background (100 mrem yr^{-1}), which is enough to cause concern about increased cancer risk but not high enough to say that this is <u>likely</u> to be lethal. It is far below the LD$_{50}$ level of 500 rad.

14-38 (a) The alpha emission of ^{239}Pu is represented $^{239}_{94}$Pu $\rightarrow$ $^{235}_{92}$U $+$ $^4_2$He.

(b) The Δm in this reaction is -0.0056267 u. Hence, $\Delta E = -5.24$ MeV, and the energy released is $+5.24$ MeV.

(c) The 1.00 g of ^{239}Pu contains 2.52×10^{21} atoms of ^{239}Pu. The decay constant is the natural logarithm of 2 divided by the half-life, or 2.876×10^{-5} yr^{-1}. The product of these two numbers is the activity: $A = 7.25 \times 10^{16}$ yr^{-1}. It is more usual to give activities in reciprocal seconds (becquerels) than in reciprocal years. Dividing by the number of seconds in a year gives $A = 2.30 \times 10^9$ s^{-1}.

(d) Use the equation for the decay of the activity $A = A_i e^{-kt}$ with $A_i = 2.30 \times 10^9$ s^{-1}, $t = 100,000$ yr, and $k = 2.876 \times 10^{-5}$ yr^{-1} to find $A = 1.30 \times 10^8$ s^{-1}. The passage of 100,000 years has reduced the activity of the plutonium to about 5.6% of its initial value.

14-40 (a) $^6_3$Li $+$ $^1_0 n$ $\rightarrow$ $^3_1$H $+$ $^4_2$He.

(b) The molar mass of the lithium from which the ^{6}Li has been depleted will be larger because ^{6}Li has a smaller atomic mass than the atomic mass of naturally abundant lithium.

14-42 The fusion reaction $2\,{}^{2}_{1}\text{H} \rightarrow {}^{4}_{2}\text{He}$ has $\Delta m = -0.0256125$ u, a result computed from the masses of the nuclides in Table 14.1 of the text. Compute the change of mass in the reaction of exactly one gram of ${}^{2}_{1}\text{H}$:

$$1 \text{ g } {}^{2}_{1}\text{H} \times \left(\frac{1 \text{ mol } {}^{2}_{1}\text{H}}{2.0141079 \text{ g } {}^{2}_{1}\text{H}} \right) \times \left(\frac{6.022 \times 10^{23} \text{ atoms}}{1 \text{ mol } {}^{2}_{1}\text{H}} \right) \times \left(\frac{-0.0256125 \text{ u}}{2 \text{ atoms}} \right)$$

$$\times \left(\frac{1.00 \times 10^{-3} \text{ kg}}{6.022 \times 10^{23} \text{ u}} \right) = -6.358 \times 10^{-6} \text{ kg}$$

This Δm is converted to kilograms so that we can multiply it by c^2 in m^2 s^{-2} and get an energy in joules; the ΔE is -5.714×10^{11} J. The energy released in the nuclear reaction is 5.714×10^{11} joules per gram of deuterium.

14-44 (a) Two equations are

$$^{231}_{92}\text{U} + {}^{0}_{-1}e^{-} \rightarrow {}^{231}_{91}\text{Pa} + \nu$$

$$^{231}_{92}\text{U} \rightarrow {}^{231}_{91}\text{Pa} + {}^{0}_{1}e^{+} + \nu$$

(b) For the first process,

$$\Delta m = m[{}^{231}_{91}\text{Pa}] - m[{}^{231}_{92}\text{U}] = 231.035881 - 231.0363 = -.0004 \text{ u}$$

For the second process,

$$\Delta m = m[{}^{231}_{91}\text{Pa}] + 2m[{}^{0}_{1}e^{+}] - m[{}^{231}_{92}\text{U}] = +0.0007 \text{ u}$$

Only the first one has $\Delta m < 0$ and is spontaneous.

14-46 Alpha decay decreases Z by 2 and A by 4; beta decay increases Z by 1 and leaves A unchanged. The atomic number of thorium is 90, the atomic number of uranium is 92, and the atomic number of lead is 82. The multi-step process of decay U $\rightarrow$ Pb emits two more alpha particles than the process Th $\rightarrow$ Pb, resulting in an A lower for the lead that derives from it. ${}^{232}_{90}\text{Th}$ loses $232 - 208 = 24$ u (6 α-particles). ${}^{238}_{92}\text{U}$ loses $238 - 206 = 32$ u (8 α-particles).

14-48

$$(2 \times 8.5 \text{ MeV}) \times 1.602177 \times 10^{-13} \text{ J MeV}^{-1} = c^2 \Delta m$$

$$= c^2 \times 0.0186234 \text{ u} \times 1.66054 \times 10^{-27} \text{ kg u}^{-1}$$

$$c^2 = 8.81 \times 10^{16} \text{ m}^2 \text{ s}^{-2}$$

$$c = 2.97 \times 10^8 \text{ m s}^{-1}$$

from Cockcroft and Walton, compared with $c = 2.9979 \times 10^8$ m s^{-1} from the measured velocity of light (agreement to 1.0%).

14-50 The decay constant for the double beta decay is 1.98×10^{-28} s^{-1}, and $N = 6.022 \times 10^{23}$ atoms. The product of these two values is the activity of the sample. It is $A = 1.2 \times 10^{-4}$ s^{-1}. Only about 10 atoms of ^{82}Se in this huge sample decay per day!

14-52 Solve the formula

$$N = \frac{A}{k} = \frac{A\, t_{\frac{1}{2}}}{\ln 2} = \frac{(3 \times 10^4 \text{ s}^{-1})(458 \times 365 \times 24 \times 60 \times 60 \text{ s})}{\ln 2}$$

$$= 6.25 \times 10^{14} \text{ atoms } {}^{241}\text{Am}$$

The mass is

$$\left(\frac{6.25 \times 10^{14} \text{ atoms}}{6.022 \times 10^{23} \text{ atoms mol}^{-1}} \right) (241 \text{ g mol}^{-1}) = 2.5 \times 10^{-7} \text{ g}$$

14-54 (a) According to problem 14-53, the disintegration rate of ^{14}C atoms in the biosphere is 0.255 Bq^{-1} g^{-1}. The overall activity of the biosphere is 1.1×10^{19} Bq. Use these as follows:

$$\frac{1.1 \times 10^{19} \text{ Bq}}{0.255 \text{ Bq g}^{-1}} = 4.3 \times 10^{19} \text{ g C}$$

(b) The carbon in the earth's crust weighs $(250 \times 10^{-6})(2.9 \times 10^{25} \text{ g}) = 7.25 \times 10^{21}$ g. The amount of carbon in the biosphere is only about 0.006 of this or about 0.6%. The rest of the carbon in the crust is tied up in rocks.

14-56 (a) Because equal numbers of ^{60}Co and ^{131}I atoms are allowed to decay for an indefinitely long time, the same number of beta-particles will be emitted by the two isotopes. The ratio of the number of rems of the two isotopes will be the same as the ratio of their energies. Taking their average energies to be the same as their maximum energies (an assumption that may not be valid) we have

$$\frac{\text{total millirems } {}^{60}\text{Co}}{\text{total millirems } {}^{131}\text{I}} = \frac{0.32 \text{ MeV}}{0.60 \text{ MeV}} = 0.53$$

(b) For a period as short as one hour, the activity of neither isotope changes significantly. Because of this and the fact that the numbers of atoms of each isotope are equal, we conclude the the number of disintegrations of a given isotope is inversely proportional to its half-life:

$$A_i = \frac{N_i \ln 2}{t_{1/2}}$$

We still have the factor of 0.53 from part (a), so that

$$\frac{\text{millirems } ^{60}\text{Co}}{\text{millirems } ^{131}\text{I}} = \left(\frac{0.32 \text{ MeV}}{0.60 \text{ MeV}}\right)\left(\frac{t_{1/2}(^{131}\text{I})}{t_{1/2}(^{60}\text{Co})}\right)$$

$$= \left(\frac{0.32}{0.60}\right)\left(\frac{8.04 \text{ days}}{5.27 \times 365 \text{ days}}\right) = 2.2 \times 10^{-3}$$

14-58 The problem can be solved by a series of unit-conversions starting from the known power of the reactor. Note the use of efficiency of the process as a conversion factor:

$$\left(\frac{1000 \times 10^6 \text{ J output}}{1 \text{ s}}\right) \times \left(\frac{100 \text{ J released}}{40 \text{ J output}}\right) \times \left(\frac{1 \text{ MeV}}{1.602 \times 10^{-13} \text{ J released}}\right)$$

$$\times \left(\frac{1 \text{ atom } ^{235}\text{U}}{200 \text{ MeV}}\right) \times \left(\frac{1 \text{ mol } ^{235}\text{U}}{6.022 \times 10^{23} \text{ atoms } ^{235}\text{U}}\right) \times \left(\frac{235 \text{ g } ^{235}\text{U}}{1 \text{ mol } ^{235}\text{U}}\right)$$

$$\times \left(\frac{3.154 \times 10^7 \text{ s}}{1 \text{ yr}}\right) = 9.6 \times 10^5 \frac{\text{g } ^{235}\text{U}}{\text{yr}}$$

14-60 The elements Li, Be, and B are not formed by the helium burning process because their most stable isotopes have atomic numbers 7, 9, and 11. Such odd–numbered isotopes require other mechanisms for their nucleosynthesis.

14-62 (a) The ratio A/Z for the elements with even Z from helium ($Z = 2$) through calcium ($Z = 20$) is generally quite near to 2. The exceptions are Be, for which the ratio is 2.25, and Ar, for which the ratio is 2.22.

(b) The expected atomic mass of argon would be about 36 based on the trend among other Z-even species.

(c) Radioactive ^{40}K decays to ^{40}Ar. This enriches naturally occurring argon in a heavy isotope relative to what would be found if there were no such decay. At the same time, it depletes potassium of its heaviest naturally occurring isotope.

14-64 The problem makes it clear that the energy emitted in the radioactive decay in 1.00 g of ^{226}Ra over the course of 1.00 hr is to be considered transferred without loss to 10.0 g of water. First, divide 1.00 g by the molar mass of the ^{226}Ra and multiply by Avogadro's number. This gives the number of atoms in the sample: 2.66×10^{21} atoms. Next, determine the activity of ^{226}Ra in disintegrations per hour

$$A = kN = \left(\frac{\ln 2}{1622 \times 365 \times 24}\right) 2.66 \times 10^{21} = 1.30 \times 10^{14} \text{ hr}^{-1}$$

Note that the half-life of ^{226}Ra is so long compared to one hour that the number of atoms remaining in the 1.00 g sample after an hour is essentially equal to the number present before. Each disintegration releases 4.79 MeV of energy. The energy released in an hour is thus the product of 4.79 MeV and the number of disintegrations per hour, or 6.23×10^{14} MeV. This energy, which equals 99.8 J, appears as heat q to be absorbed by the water. Recalling that $q = c_s m \Delta T$ and substituting $m = 10.0$ g and $c_s = 4.18$ J K^{-1}g^{-1} allows determination of the change in temperature, $\Delta T = 2.39$ K. The final temperature of the water is 27.4°C.

14-66 (a) Take the half equation

$$4 \, H_3O^+(aq) + 4 \, e^- \rightarrow 2 \, H_2(g) + 4 \, H_2O(l)$$

and subtract the half equation given, learning

$$Zr(s) + 2 \, H_2O(l) \rightarrow ZrO_2(s) + 2 \, H_2(g)$$

The $\Delta \mathcal{E}°$ is 0 - (-1.43 V) = 1.43 V > 0 so the proposed reaction is spontaneous.

(b)

$$\log_{10} K = \frac{n}{0.0592 \, \text{V}} \Delta \mathcal{E}° = \frac{4(1.43 \, \text{V})}{0.0592 \, \text{V}} = 96.6$$

$$K = 4 \times 10^{96}$$

(c) The large equilibrium constant from part (b) means that if zirconium fuel rods come into contact with water there is a thermodynamic tendency toward formation of hydrogen.

Chapter 15

Quantum Mechanics and Atomic Structure

15-2 The wavelength of this chemical wave is 1.2 cm; its frequency is 1/42 s^{-1}. The speed of propagation is the product of these two values, 0.029 cm s^{-1}.

15-4 Substitute in the equation $c = \lambda\nu$. Assume that the gamma rays are propagating through a vacuum so $c = 2.9979 \times 10^8$ m s^{-1}. The wavelength is then

$$\lambda = \frac{c}{\nu} = \frac{2.9979 \times 10^8 \text{ m s}^{-1}}{2.83 \times 10^{20} \text{ s}^{-1}} = 1.06 \times 10^{-12} \text{ m}$$

This equals 0.0106 Å.

15-6 (a) Substituting in $c = \lambda\nu$ gives the frequency of 488 nm light to be 6.14×10^{14} s^{-1}.

(b) We divide twice the distance from the earth to the moon by the speed of light. The answer is 2.5 s.

15-8 The wavelength of the ultrasonic wave is its speed of propagation divided by its frequency. It comes out to 0.030 m, or 3.0 cm. The resolution in the sonic image is never better than the wavelength of sound used. If lower frequency ($\nu = 8000$ s^{-1}) is used, then the wavelength is longer (18.7 cm) and all detail in the image of the fetus is lost.

15-10 The red light has lower frequency and hence lower energy than the green light. If green light ejects no electrons from the copper surface, then red light also ejects no electrons.

15-12 The maximum wavelength of light is the wavelength that supplies photons just energetic enough to overcome the work function of the surface:

$$\lambda = \frac{c}{\nu} = \frac{hc}{E} = \frac{(6.626 \times 10^{-34} \text{J s})(2.9979 \times 10^8 \text{ m s}^{-1})}{4.41 \times 10^{-19} \text{ J}} = 4.50 \times 10^{-7} \text{ m}$$

15-14 The wavelength of the radiation must be short enough to make the photon energetic enough to eject an electron from the surface of the tungsten. The work function is 7.29×10^{-19} J, which is supplied by photons of wavelength 272 nm or shorter (computed using $\lambda = hc/E$). If the ejected electron has a velocity of 2.00×10^6 m s^{-1}, its kinetic energy is 1.82×10^{-18} J, by application of the formula $K.E. = 1/2\,mv^2$ with $m = 9.109 \times 10^{-31}$ kg and v as given. The energy required from the photon is the work function plus this kinetic energy; the sum equals 2.55×10^{-18} J. The corresponding λ is 77.9 nm.

15-16 A wavelength of 454 nm is in the blue region of the spectrum.

15-18 The energy lost by the potassium atom is carried away by a photon. The wavelength of the photon is

$$\lambda = \frac{hc}{\Delta E} = \frac{(6.626 \times 10^{-34} \text{J s})(2.9979 \times 10^8 \text{ m s}^{-1})}{4.9 \times 10^{-19} \text{ J}} = 4.1 \times 10^{-7} \text{ m} = 410 \text{ nm}$$

This wavelength is in the violet region of the spectrum.

15-20 (a) The energy carried by a photon is the product of Planck's constant and its frequency $E = h\nu$. In this case $E = 3.8 \times 10^{-19}$ J.

(b) The power of the laser is 10 W, which is 10 J s^{-1}. Hence:

$$\frac{10 \text{ J}}{1 \text{ s}} \times \left(\frac{1 \text{ photon}}{3.82 \times 10^{-19} \text{ J}} \right) = 2.6 \times 10^{19} \frac{\text{photon}}{\text{s}}$$

15-22 The three emission lines connect each possible pair of levels (see diagram).

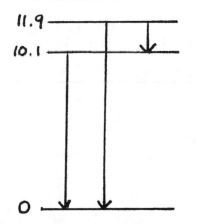

The corresponding wavelengths of emitted light are given by

$$\lambda = \frac{hc}{\Delta E} = \frac{12.3982 \times 10^{-7} \text{ m}}{V_{\text{thr}}[\text{V}]}$$

(recall Example 15.3). Substituting $V_{\text{thr}} = 10.1$, 11.9, and 1.8 V (the last of these is the voltage difference between the two excited states) gives wavelengths of

$$\lambda = 1.23 \times 10^{-7} \text{ m}, \ 1.04 \times 10^{-7} \text{ m}, \text{ and } 6.9 \times 10^{-7} \text{ m}$$

$$\text{or } 1230 \text{ Å}, \ 1040 \text{ Å}, \text{ and } 6900 \text{ Å}$$

15-24 (a) The energy change to remove an electron from a ground-state atom is

$$\Delta E = 13.6 \text{ eV} \left(\frac{1.602 \times 10^{-19} \text{ J}}{1 \text{ eV}} \right)$$

$$= 2.179 \times 10^{-18} \text{ J}$$

The wavelength to ionize it is

$$\lambda = \frac{c}{\nu} = \frac{hc}{h\nu} = \frac{hc}{\Delta E}$$

$$= \frac{(6.626 \times 10^{-34} \text{ J s})(2.9979 \times 10^8 \text{ m s}^{-1})}{2.179 \times 10^{-18} \text{ J}}$$

$$= 9.12 \times 10^{-8} \text{ m} = 912 \text{ Å}$$

(b) $\frac{1}{2}mv^2 = \Delta E$

$$v^2 = \frac{2\Delta E}{m} = \frac{2(2.179 \times 10^{-18} \text{ J})}{9.109 \times 10^{-31} \text{ kg}} = 4.78 \times 10^{12} \text{ m}^2 \text{ s}^{-2}$$

$$v = 2.19 \times 10^6 \text{ m s}^{-1}$$

Converting to miles per hour gives

$$v = 2.19 \times 10^6 \text{ m s}^{-1} \left(\frac{1 \text{ ft}}{0.3048 \text{ m}} \right) \left(\frac{1 \text{ mile}}{5280 \text{ ft}} \right) \left(\frac{3600 \text{ s}}{1 \text{ hr}} \right)$$

$$= 4.89 \times 10^6 \text{ miles hour}^{-1}$$

(c) For thermal excitation, $k_{\text{B}}T \approx \Delta E$

$$T \approx \frac{\Delta E}{k_{\text{B}}} = \frac{2.179 \times 10^{-18} \text{ J}}{1.381 \times 10^{-23} \text{ J K}^{-1}} = 1.58 \times 10^5 \text{ K}$$

15-26 According to the Bohr model, the radius of a one-electron atom or ion is

$$r = \frac{n^2}{Z}a_o = \frac{n^2}{Z}(5.29 \times 10^{-11} \text{ m})$$

Substitution of $Z = 2$ for helium and $n = 5$ gives $r = 6.61 \times 10^{-10}$ m. The energy of any state of a one-electron atom or ion is given by

$$E = -\frac{Z^2}{n^2}(2.18 \times 10^{-18} \text{ J})$$

In the case of a He^+ ion in the $n = 5$ state, this energy equals -3.49×10^{-19} J. Removing the electron means changing the energy of the atom to $E = 0$. The change in energy of one atom is this final value minus the initial value, or $+3.49 \times 10^{-19}$ J. For a mole of atoms the energy change is Avogadro's number times larger or 210 kJ. The energy of the He^+ ion in the $n = 3$ state is -9.69×10^{-19} J. The change in energy of the ion in the $5 \rightarrow 3$ transition is the $n = 3$ energy (the final energy) minus the $n = 5$ energy (the initial energy). Hence $\Delta E = -6.20 \times 10^{-19}$ J. The transition gives off energy, as shown by its negative ΔE. The frequency of the photon that carries this energy away is 9.36×10^{14} s^{-1}, and the wavelength is 320 nm.

15-28 In the spectrum of Be^{3+} (a $Z = 4$ ion) the series of lines analogous to the Lyman series of atomic hydrogen has wavelengths equal to 1/16 of the wavelengths of the Lyman series; the frequencies in this series are 16 times the frequencies in the Lyman series. A similar scaling occurs with the series analogous to the Balmer series. These conclusions follow from the dependence of the energy of the states of hydrogen-like ions on Z^2. Thus, for the frequencies in the Lyman-like series

$$\nu = (16) \times 3.29 \times 10^{15} \left(\frac{1}{1^2} - \frac{1}{n_{\text{final}}^2} \right) \text{ s}^{-1}$$

The ν's for the first three n_{final}'s are 3.95×10^{16}, 4.68×10^{16}, and 4.94×10^{16} s^{-1}. The corresponding wavelengths are 7.59, 6.41, and 6.07 nm, in the x-ray region.

For the frequencies in the Balmer-like series

$$\nu = (16) \times 3.29 \times 10^{15} \left(\frac{1}{2^2} - \frac{1}{n_{\text{final}}^2} \right) \text{ s}^{-1}$$

The ν's for the first three n_{final}'s are 7.31×10^{15}, 9.87×10^{15}, and 1.11×10^{16} s^{-1}. The corresponding wavelengths are 41.0, 30.4, and 27.1 nm, in the ultraviolet region.

15-30 (a) In the ground state of the standing wave, there is one half-wavelength along the bond, L. Hence $\lambda_1 = 2.0$ Å. In the first excited state, $n = 2$; there are two half-wavelengths along the bond, so $\lambda_2 = L = 1.0$ Å.

(b) The number of nodes is one less than the quantum number describing the standing wave. Hence, there is 1 node.

15-32 (a) We know the mass of the electron so we can calculate the v of the electrons from their kinetic energy, from the relationship $K.E. = 1/2\, m_e v^2$. It is 6.614×10^6 m s^{-1}. The wavelength is 0.110 nm (using $\lambda = h/p = h/m_e v$).

(b) The helium atom moving at 353 m s^{-1} gives $\lambda = 0.282$ nm.

(c) The krypton atom moving at 299 m s^{-1} has $\lambda = 0.0159$ nm.

15-34 (a) We use the Heisenberg uncertainty principle:

$$\Delta x \cdot \Delta(mv) \geq h/4\pi$$

with $m = 9.109 \times 10^{-31}$ kg and $v = 3.0 \times 10^8$ m s^{-1}. We find $\Delta x \geq 1.93 \times 10^{-13}$ m. This means that if we know nothing about the speed an electron, we can know its location to at best ± 0.002 Å.

(b) Because the helium atom is much more massive its Δx is much smaller. By a similar computation it is 2.65×10^{-17} m.

15-36

$$\Delta E = \frac{hc}{\lambda} = \frac{6.626 \times 10^{-34} \text{ J s} \times 2.9979 \times 10^8 \text{ m s}^{-1}}{800 \times 10^{-9} \text{ m}} = 2.483 \times 10^{-19} \text{ J}$$

$$\Delta E = \frac{h^2}{8\, mL^2}\left[(2)^2 + (1)^2 + (1)^2 - (1)^2 - (1)^2 - (1)^2\right] = \frac{3h^2}{8\, mL^2}$$

$$L^2 = \frac{3 \times (6.626 \times 10^{-34} \text{ J s})^2}{8 \times 9.109 \times 10^{-31} \text{ kg} \times 2.483 \times 10^{-19} \text{ J}} = 7.279 \times 10^{-19} \text{ m}^2$$

$$L = 8.53 \times 10^{-10} \text{ m} = 8.53 \text{ Å}$$

15-38 (a) This combination is not allowed because m_s is never equal to zero. If m_s were changed to $\pm 1/2$, this combination would be allowed.

(b) This combination is allowed. It specifies a 2s-electron.

(c) This combination is allowed. It specifies a 7d-electron.

(d) This combination is not allowed. The quantum number l is never negative.

15-40 (a) 3d   (b) 7g   (c) 5p.

15-42 (a) A $3d$-orbital has 0 radial and 2 angular nodes; (b) a $7g$-orbital has 2 radial and 4 angular nodes; (c) a $5p$-orbital has 3 radial and 1 angular node.

15-44 (a)

$$R(3p) = \frac{4}{81\sqrt{6}} \left(\frac{Z}{a_0}\right)^{3/2} (6\sigma - \sigma^2) \exp\left(-\frac{\sigma}{3}\right)$$

A node occures at $R(3p) = 0$, where

$$6\sigma - \sigma^2 = 0$$

$$\sigma = 6 = \frac{r}{0.529 \times 10^{-10} \text{ m}}$$

$$r = 6 \times 0.529 \times 10^{-10} \text{ m} = 3.17 \times 10^{-10} = 3.17 \text{ Å}$$

(b)

$$R(3s) = \frac{2}{81\sqrt{3}} \left(\frac{Z}{a_0}\right)^{3/2} (27 - 18\sigma + 2\sigma^2) \exp\left(-\frac{\sigma}{3}\right)$$

Nodes occur at the roots of $27 - 18\sigma + 2\sigma^2 = 0$

$$\frac{r}{a_0} = \sigma = \frac{18 \pm \sqrt{324 - 8 \times 27}}{4} = 1.902, 7.098$$

$$r_1 = 1.902 \times 0.529 \times 10^{-10} \text{ m} = 1.01 \text{ Å}$$

$$r_2 = 7.098 \times 0.529 \times 10^{-10} \text{ m} = 3.75 \text{ Å}$$

15-46 (a) P: $[\text{Ne}]3s^2 3p^3$ (b) Tc: $[\text{Kr}]4d^5 5s^2$ (c) Ho: $[\text{Xe}]4f^{11}6s^2$

15-48 Li^- $(1s^2 2s^2)$ is diamagnetic; B^+ $(1s^2 2s^2)$ is diamagnetic;
F^- $(1s^2 2s^2 2p^6)$ is diamagnetic; Al^{3+} $(1s^2 2s^2 2p^6)$ is diamagnetic.
All of the preceding have an even number of electrons and all electrons paired.
S^- $(1s^2 2s^2 2p^6 3s^2 3p^5)$ is paramagnetic.
Ar^+ has the same ground-state configuration as S^- and is therefore also paramagnetic.
Br^+ $([\text{Ar}]3d^{10}4s^2 4p^4)$ is paramagnetic; Te^- $([\text{Kr}]4d^{10}5s^2 5p^5)$ is paramagnetic.
A species with an odd number of electrons must be paramagnetic; species with an even number of electrons may be diamagnetic or paramagnetic.

15-50 (a) The species is an atom and is therefore electrically neutral. It has 76 electrons. The element with $Z = 76$ is osmium. (b) The ion is F^-. (c) The ion is Ag^{5+}.

15-52 (a) This element would follow the completion of the $n = 7$ row in the periodic table and would have $Z = 119$.

(b) We note that 137 exceeds 119 by 18. This difference would be neatly explained by the intermediate filling of nine $5g$ orbitals by 18 electrons before the closing of the seventh row of the periodic table.

15-54 The three noble-gas atoms would have $Z = 3$ (corresponding to "$1s^3$"), $Z = 15$ ("$1s^3 2s^3 2p^9$ "), and $Z = 27$ ("$1s^3 2s^3 2p^9 3s^3 3p^9$ ").

15-56 (a) Sm (b) Ca (c) I⁻ (d) Ge (e) Rb.

15-58 (a) The two are isoelectronic, and S^{2-} has a lesser nuclear charge; it is larger. (b) The Tl^+ is larger. Loss of two electrons to give Tl^{3+} reduces electron-electron repulsions and so allows contraction. (c) The Ce^{3+} ion is larger, considering the lanthanide contraction. (d) The I⁻ ion is larger; its outer electrons are in the $n = 5$ shell.

15-60 (a) Xe should have a higher IE_1 than Bi because of its closed-shell configuration.

(b) Selenium (Se) should have a higher IE_1 than Te; the two are in the same group and Se is higher up the periodic table.

(c) Yttrium (Y) should have a higher IE_1 than Rb. The two are in the same row and Y is farther to the right.

(d) Neon (Ne) should have a higher IE_1 than K since its IE_1 exceeds that of Ar, which definitely has a higher IE_1 than K.

15-62 The reasoning uses the patterns in the periodic table as in problem 15-60:

(a) Rb (b) I (c) Te (d) Cl.

15-64 Calcium does have a positive affinity for an electron, but the quoted number is very small. We convert it from kJ mol⁻¹ to J per atom and then use $E = hc/\lambda$ to compute the wavelength. Using the most recent value for the EA of Ca(2.0 kJ mol⁻¹) gives 6.0×10^{-5} m; using the earlier value of 1.7 kJ mol⁻¹ gives 7.0×10^{-5} m. Infrared radiation is energetic enough to remove the extra electron from Ca⁻.

15-66 The time it takes for any electromagnetic waves to arrive from Cygnus A is its distance divided by the speed of light. This is 3×10^{24} m divided by 3.00×10^8 m s⁻¹ or 10^{16} s, which equals 3×10^8 years. In other words, Cygnus A is 300 million light-years away.

The frequency of the radio wave is c divided by its wave length. It is 3.0×10^7 s⁻¹.

15-68

$$\Delta E_{\text{x-ray}} = \frac{hc}{\lambda} = \frac{6.626 \times 10^{-34} \text{ J s} \times 3.00 \times 10^8 \text{ m s}^{-1}}{0.20 \times 10^{-9} \text{ m}} = 9.9 \times 10^{-16} \text{ J}$$

$$\Delta E_{\text{AM radio}} = \frac{6.626 \times 10^{-34} \text{ J s} \times 3.00 \times 10^8 \text{ m s}^{-1}}{200 \text{ m}} = 9.9 \times 10^{-28} \text{ J}$$

The x-ray photon is capable of inducing a chemical reaction for which the required energy is 6.0×10^5 kJ mol^{-1}. (This exceeds the dissociation energy of every chemical substance) The AM photon has an energy of only 6.0×10^{-4} J mol^{-1} and is ineffective in influencing chemical reactions.

15-70

$$v = \frac{nh}{2\pi m_e r}$$

$$r = \frac{\epsilon_0 n^2 h^2}{\pi Z e^2 m_e}$$

$$v = \frac{\pi Z n h e^2 m_e}{2\pi m_e \epsilon_0 n^2 h^2} = \frac{Z e^2}{2\epsilon_0 n h}$$

$$v_e(\text{He}^+) = \frac{2 \times (1.60218 \times 10^{-19} \text{ C})^2}{2 \times 8.8542 \times 10^{-12} \text{ C}^2 \text{ J}^{-1} \text{ m}^{-1} \times 1 \times 6.626 \times 10^{-34} \text{ J s}}$$

$$= 4.38 \times 10^6 \text{ m s}^{-1}$$

$$v_e(\text{U}^{91+}) = \frac{92 \times (1.60218 \times 10^{-19} \text{ C})^2}{2 \times 8.8542 \times 10^{-12} \text{ C}^2 \text{ J}^{-1} \text{ m}^{-1} \times 1 \times 6.626 \times 10^{-34} \text{ J s}}$$

$$= 2.01 \times 10^8 \text{ m s}^{-1}$$

The velocity of light is 2.9979×10^8 m s^{-1}. Relativistic effects will be very important in U^{91+}, but less important in He$^+$.

15-72 The C^{5+} ion is a hydrogen-like ion with $Z = 6$. All transitions in its spectrum have frequencies that fit the following formula with $Z = 5$ and integral n:

$$\nu = \frac{c}{\lambda} = -Z^2(3.29 \times 10^{15}) \left(\frac{1}{n_{\text{initial}}^2} - \frac{1}{n_{\text{final}}^2} \right)$$

Suppose that green light has a wavelength range running from 500 to 550 nm (Fig. 15.3). Insert these wavelengths expressed in meters, $Z = 6$, and the various constants in SI units into the formula. Then, all units cancel out and

$$\left(\frac{1}{n_{\text{final}}^2} - \frac{1}{n_{\text{initial}}^2} \right) = 0.00460 \text{ to } 0.00506$$

Now, systematically try combinations of integers that give the desired result. Note that n_{final} must be less than $n_{initial}$. The first combination that works is $n_{initial} = 8$ and $n_{final} = 7$ for which

$$\left(\frac{1}{7^2} - \frac{1}{8^2}\right) = 0.00478$$

15-74 (a) The loss in intensity of the tone is due to destructive interference between the two loudspeakers superseding constructive interference as one loudspeaker is moved closer Destructive interference occurs when one wave-train lags the other by $1, 2, 3 \ldots$ half-wavelengths. The half-wavelength $1/2\lambda$ is thus 0.16 (or 0.080 or 0.0533...) m, and λ is 0.32 m. We take this first answer because losses of intensity of the tone would presumably have been reported at movement distances less than 0.16 m if the wavelength were shorter.

(b) The frequency is the speed of the sound divided by its wavelength. It is 1.1×10^3 s^{-1}.

15-76 (a)
$$\lambda = \frac{h}{mv} = \frac{1 \text{ J s}}{0.145 \text{ kg} \times 20 \text{ m s}^{-1}} = 0.34 \text{ m}$$

(b)
$$\Delta p = 0.145 \text{ kg} \times 2 \text{ m s}^{-1} = 0.29 \text{ kg m s}^{-1}$$
$$\Delta p \Delta x \geq \frac{h}{4\pi} = \frac{1 \text{ J s}}{4\pi}$$
$$\Delta x \geq \frac{1 \text{ J s}}{0.29 \times 4 \times \pi} = 0.27 \text{ m}$$

(c)
$$a_0 = \frac{\epsilon_0 h^2}{\pi e^2 m_e} = \frac{8.554 \times 10^{-12} \text{ C J}^{-1} \text{ m}^{-1} (1 \text{ J s})^2}{\pi \times (1 \text{ C})^2 \times 0.001 \text{ kg}} = 2.82 \times 10^{-9} \text{ m} = 28 \text{ Å}$$

15-78
$$E_{n_x n_y n_z} = \frac{h^2}{8m}\left[\frac{n_x^2}{(2L)^2} + \frac{n_y^2}{L^2} + \frac{n_z^2}{L^2}\right]$$

Ground state:
$$E_{111} = \frac{h^2}{8\,mL^2}(1/4 + 1 + 1) = \frac{9\,h^2}{32mL^2}$$

Excited states:
$$E_{211} = \frac{h^2}{8\,mL^2}(4/4 + 1 + 1) = \frac{3\,h^2}{8mL^2}$$

$$E_{311} = \frac{h^2}{8\,mL^2}(9/4 + 1 + 1) = \frac{17\,h^2}{32mL^2}$$

$$E_{112} = E_{121} = \frac{h^2}{8\,mL^2}(1/4 + 4 + 1) = \frac{21\,h^2}{32mL^2}$$

$$E_{411} = E_{221} = E_{212} = \frac{h^2}{8\,mL^2}(6) = \frac{3\,h^2}{4mL^2}$$

$$E_{321} = E_{312} = \frac{h^2}{8\,mL^2}(9/4 + 4 + 1) = \frac{29\,h^2}{32mL^2}$$

15-80 (a) The probability density of finding an electron in the vicinity of any point is equal to the square of the wave function of the electron evaluated at that point. Thus, the probability density of finding the 1s electron of the H atom at a distance r from the nucleus is:

$$\psi^2 = (1/\pi a_0^3)\exp(-2r/a_0)$$

The portion in the first parentheses equals 2.15×10^{30} m^{-3} and the exponential term is unitless. At r equals 0 the exponential term equals 1 and ψ^2 is 2.15×10^{30} m^{-3}.

The probability of finding the electron *exactly at* a mathematical point is zero because a point has no volume to accommodate the electron. The small sphere centered at the nucleus has however a volume of 1 pm^3 (1.0×10^{-36} m^3). Over the very short distance between the center and surface of this sphere ψ^2 stays nearly constant at 2.15×10^{30} m^{-3}. It follows that the probability of finding the electron within the small sphere is about:

$$p = 2.15 \times 10^{30} \text{ m}^{-3} \times 1.00 \times 10^{-36} \text{ m}^3 = 2.15 \times 10^{-6}$$

(b) At a distance a_0 or 52.9 pm (0.529×10^{-10} m) from the nucleus, the value of the functions ψ^2 is *less* than it is at the nucleus. The exponential part of the function drops rapidly as r increases:

$$\psi^2(\text{at } a_0) = (2.15 \times 10^{30} \text{ m}^{-3})e^{-2r/a_0} = 2.91 \times 10^{29} \text{ m}^{-3}$$

Assume that ψ^2 is constant throughout the 1 pm^3 volume which the problem specifies. The chance of finding the electron at 52.9 pm in a fixed direction is:

$$p = 2.91 \times 10^{29} \text{ m}^{-3} \times 1.0 \times 10^{-36} \text{ m}^3 = 2.91 \times 10^{-7}$$

(c) A spherical shell of thickness 1 pm and radius 52.9 pm has a volume:

$$V_{shell} = 4\pi r^2 \Delta r = 4\pi(52.9 \text{ pm})^2 \times 1 \text{ pm} = 3.52 \times 10^{-32} \text{ m}^3$$

This substantially larger volume naturally has a greater probability of holding the electron than does the tiny 1 pm^3 volume element considered in part b:

$$p = 2.91 \times 10^{29} \text{ m}^{-3} \times 3.52 \times 10^{-32} \text{ m}^3 = 0.0102$$

15-82 (a)

$$Y(p_x) = \left(\frac{3}{4\pi}\right)^{1/2} \sin\theta \cos\varphi$$

$$Y(p_y) = \left(\frac{3}{4\pi}\right)^{1/2} \sin\theta \sin\varphi$$

$$Y(p_z) = \left(\frac{3}{4\pi}\right)^{1/2} \cos\theta$$

The radial part of the wave function need not be taken into account because it does not affect the angular symmetry.

$$Y^2(p_x) + Y^2(p_y) + Y^2(p_z) = \left(\frac{3}{4\pi}\right)(\sin^2\theta\cos^2\varphi + \sin^2\theta\sin^2\varphi + \cos^2\theta)$$

$$= \left(\frac{3}{4\pi}\right)(\sin^2\theta(\cos^2\varphi + \sin^2\varphi) + \cos^2\theta) = \left(\frac{3}{4\pi}\right)$$

Hence, the N atom is spherically symmetric because ψ^2 is independent of angle.

(b) The following species are spherically symmetric: F^-, Na, S^{2-}, Cu, Mo, Sb, Au.

15-84

Z		$r(\text{Å})$	
57		1.87	
58		1.82	
59		1.82	
60		1.81	
61		1.81	
62		1.80	
63	(Eu)	2.00	←
64		1.79	
65		1.76	
66		1.75	
67		1.74	
68		1.73	
69		1.72	
70	(Yb)	1.94	←
71		1.72	

(a) The trend in atomic radius with Z for the rare earth elements is generally downward (the so-called lanthanide contraction). It is caused by the increasing nuclear charge that attracts the electrons in the $4f$ sub-shell toward the nucleus.

(b) The elements 63(Eu) and 70(Yb) are the exceptions to the trend. They contain, respectivly, a half-filled and a filled $4f$ sub-shell.

15-86 The third ionization energy of Li is the ΔE for the process $Li^{2+}(g) \rightarrow Li^{3+}(g) + e^-$. The ejection of a $1s$-electron from a lithium atom on the other hand is represented $Li(g) \rightarrow Li^{+*}(g) + e^-$. where the star indicates that the Li^+ ion is in an excited state. In both cases a $1s$-electron is removed. The IE_3 is larger because the $1s$-electron is removed against the full attraction of the $Z = 3$ nucleus whereas the removal of the $1s$-electron in the photoelectron spectroscopy experiment is from a $Z = 3$ nucleus but assisted by repulsions from the other two electrons.

15-88 The first ionization energy of K is 419 kJ mol^{-1}

$$\Delta E_1 = \frac{hc}{\lambda_1} = \frac{6.626 \times 10^{-34} \text{J s} \times 2.9979 \times 10^8 \text{ m s}^{-1} \times 6.022 \times 10^{23} \text{ mol}^{-1}}{650 \times 10^{-9} \text{ m}}$$

$$= 1.84 \times 10^5 \text{ J mol}^{-1} = 184 \text{ kJ mol}^{-1}$$

$$419 = 184 + \Delta E_2; \quad \Delta E_2 = 235 \text{ kJ mol}^{-1}$$

$$\lambda_2 = \frac{hc}{\Delta E_2} = \frac{6.626 \times 10^{-34} \text{ J s} \times 3.00 \times 10^8 \text{ m s}^{-1} \times 6.022 \times 10^{23} \text{ mol}^{-1}}{2.35 \times 10^5 \text{ J mol}^{-1}}$$

$$= 509 \times 10^{-9} \text{ m}$$

This is the maximum wavelength the second photon may have.

15-90 The energy of each photon is

$$E = h\nu = (6.626 \times 10^{-34} \text{ J s})(2.45 \times 10^9 \text{ s}^{-1}) = 1.62 \times 10^{-24} \text{ J per photon}$$

The energy change to heat the water is the heat absorbed

$$\Delta E = q = (4.2 \text{ J K}^{-1} \text{ g}^{-1})(100.0 \text{ g})(100 - 15 \text{ K}) = 3.57 \times 10^4 \text{ J}$$

The number of photons is the quotient of these two numbers:

$$\frac{3.57 \times 10^4 \text{ J}}{1.62 \times 10^{-24} \text{ J per photon}} = 2.2 \times 10^{28} \text{ photons}$$

15-92 (a) If an electron is confined within a nucleus with a radius of 1×10^{-15} m, then the uncertainty in its position is at most 2×10^{-15} m. If Δx exceeded this, the diameter of the nucleus, the electron could hardly be said to be confined. The Heisenberg uncertainty principle states:

$$\Delta p \Delta x > h/4\pi$$

The right-hand side of this inequality is 5.27×10^{-35} J s. It follows that Δp_{min}, the *minimum* uncertainty in the momentum of the confined electron, is 2.64×10^{-20} kg m s^{-1}. If it is supposed that the mass of the electron is its rest mass, 9.11×10^{-31}kg, then, using the definition of momentum, $p = mv$:

$$\Delta v_{min} = \Delta p_{min}/m = 2.9 \times 10^{10} \text{ m s}^{-1}$$

The minimum uncertainty in the velocity of the confined electron exceeds the speed of light. Because this is impossible, the electron cannot be confined in a nucleus.

(b) The minimum uncertainty in the momentum of a proton or neutron confined in the nucleus is again 2.64×10^{-20} kg m s^{-1}. The mass of a proton or neutron (about 1.67×10^{-27} kg) is greater than the mass of an electron. Hence, the minimum uncertainty in the velocity of the confined proton or neutron is 1.6×10^{7} m s^{-1}, which is only 5 percent of the speed of light. The proton and neutron can be confined to the nucleus.

Chapter 16

Quantum Mechanics and Molecular Structure

16-2 (a) The correlation diagram for O_2^- is identical to that given for O_2 in text Figure 16.6b except that an additional electron is added to either the $\pi_{2p_x}^*$ or $\pi_{2p_y}^*$ level.

(b) The ground-state electron configuration of O_2 is listed in Table 16.2. The configurations of the ions in the problem can be derived by removal of electrons from the highest occupied or addition to the lowest unoccupied molecular orbitals:

O_2^+: $(\sigma_{2s})^2(\sigma_{2s}^*)^2(\sigma_{2p_z})^2(\pi_{2p})^4(\pi_{2p}^*)^1$

O_2: $(\sigma_{2s})^2(\sigma_{2s}^*)^2(\sigma_{2p_z})^2(\pi_{2p})^4(\pi_{2p}^*)^2$

O_2^-: $(\sigma_{2s})^2(\sigma_{2s}^*)^2(\sigma_{2p_z})^2(\pi_{2p})^4(\pi_{2p}^*)^3$

O_2^{2-}: $(\sigma_{2s})^2(\sigma_{2s}^*)^2(\sigma_{2p_z})^2(\pi_{2p})^4(\pi_{2p}^*)^4$

(c) The bond orders of the species are 5/2 (for O_2^+), 2 (for O_2), 3/2 (for O_2^-) and 1 (for O_2^{2-}).

(d) All of the species except O_2^{2-} should be paramagnetic. A species with an odd number of electrons is automatically paramagnetic; the reason for the paramagnetism of ordinary O_2 is discussed in the text.

(e) In the series given in (c), each additional electron occupies a π^* antibonding orbital. The bond dissociation energy decreases along the series.

16-4 It is not necessary to show any configuration for core electrons, which play little part in bonding. Then:

I_2: $(\sigma_{5s})^2(\sigma_{5s}^*)^2(\sigma_{5p_z})^2(\pi_{5p})^4(\pi_{5p}^*)^4$ This substance is diamagnetic.

16-6 (a) X_2: $(\sigma_{2s})^2(\sigma_{2s}^*)^2(\sigma_{2p})^2(\pi_{2p})^4(\pi_{2p}^*)^2$: O_2. Bond order: 2.

(b) Q_2^-: $(\sigma_{2s})^2(\sigma_{2s}^*)^2(\pi_{2p})^3$: B_2^-. Bond order: $\frac{3}{2}$.

(c) Z_2^{2+}: $(\sigma_{2s})^2(\sigma_{2s}^*)^2(\sigma_{2p})^2(\pi_{2p})^4(\pi_{2p}^*)^2$: F_2^{2+}. Bond order: 2.

16-8 (a) paramagnetic   (b) paramagnetic   (c) paramagnetic

16-10 The nitrogen atomic orbitals should be lower in energy than the corresponding beryllium orbitals. The molecule of BeN has 7 valence electrons and a bond order of 3/2. The substance should be paramagnetic.

16-12 The nitrosyl molecular ion forms from nitrogen oxide by the loss of an electron from the highest occupied molecular orbital. The electron configuration of NO is $(\sigma_{2s})^2(\sigma_{2s}^*)^2(\pi_{2p})^4(\sigma_{2p_z})^2(\pi_{2p}^*)^1$ so the electron comes from a π_{2p}^* orbital. Because this is an antibonding orbital, the bonding in NO^+ should be stronger and the bond should be shorter than in NO. Moreover, NO^+ should be diamagnetic.

16-14 The HeH^+ ion has the electron configuration $(\sigma_{1s})^2$. Its bond order is 1, and it is diamagnetic. The lower energy state should be reached by the reaction $HeH^+ \rightarrow He + H^+$. This set of products is more stable than $He^+ + H$ because in it the two electrons are both close to the +2 charge of the helium nucleus instead of the +1 charge of the hydrogen nucleus and very roughly the same distance from each other.

16-16 The central O in the H_3O^+ has $SN = 4$. We expect sp^3 hybridization on the O. The molecular ion will be pyramidal.

16-18 The Lewis structures are

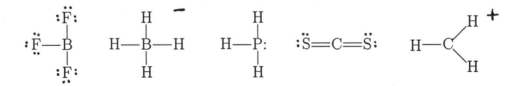

(a) The central B in BF_3 has $SN = 3$. It is sp^2 hybridized, and the molecule is trigonal-planar.
(b) The central B in BH_4^- has $SN = 4$. It is sp^3 hybridized, and the molecular ion is tetrahedral.
(c) The central P in PH_3 has $SN = 4$. It is sp^3 hybridized, and the molecule is pyramidal.
(d) The central C in CS_2 has $SN = 2$. It is sp hybridized, and the molecule is linear.

(e) The central C in CH_3^+ has $SN = 3$. It is sp^2 hybridized, and the molecular ion is trigonal-planar.

16-20 The tetrahedral ClO_4^- ion uses sp^3 hybrid orbitals on the central Cl (which has SN 4). The pyramidal ClO_3^- also uses sp^3 hybrid orbitals on the central Cl, which has SN 4.

16-22 The carbon atom in $N{\equiv}C{-}Cl$ has a steric number of 2. It is sp hybridized, and the molecule should be linear.

16-24 For NO_2^+, sp hybrid orbitals on central N atom, forming σ-bonds with p_z orbitals on outer oxygen atoms (4 electrons).

Lone pair $2s$ orbitals on outer oxygens (4 electrons).

π_x, π_y, π_x^{nb} and π_y^{nb} each with 2 electrons.

Linear and not paramagnetic.

For NO_2 and NO_2^-, sp^2 hybrid orbitals on central N atom, two forming σ-bonds with outer $2p$ orbitals and one with a lone pair on N (6 electrons).

Lone pair $2s$ and $2p$ orbitals on outer atoms (8 electrons).

π orbital containing 1 (NO_2) or 2(NO_2^-) electrons.

Both are non-linear, but only NO_2 is paramagnetic.

16-26 The nitrate ion has 24 valence electrons. Three resonance structures are needed to represent the equivalence of the three N to O bonds:

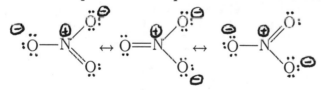

The SN of the central nitrogen is 3, which in VSEPR theory predicts trigonal-planar geometry about the central nitrogen. This corresponds to sp^2 hybridization of the valence orbitals of the central N. Six valence electrons are shared between the N and the O's in these orbitals. The remaining $2p$-orbital of the nitrogen combines with one $2p$-orbital on each of the three oxygen atoms to form a set of four π-molecular orbitals. The lowest lying of these, a bonding orbital, is occupied by a pair of electrons. The net effect is that the four atoms are bonded by eight electrons: each of the three links between N and O has bond order 4/3.

16-28 The energy spacing is

$$\frac{h^2}{8\pi^2 I}[1(2) - 0(1)] = \frac{h^2}{4\pi^2 I}$$

$$\mu = \frac{(1.007825)(18.9984)\text{ u}}{1.007825 + 18.9984} \times \left(1.66054 \times 10^{-27}\text{ kg u}^{-1}\right)$$

$$= 1.589 \times 10^{-27}\text{ kg}$$

$$I = \mu R_e^2 = \left(1.589 \times 10^{-27}\text{ kg}\right)\left(0.926 \times 10^{-10}\text{ m}\right)^2$$

$$= 1.363 \times 10^{-47}\text{ kg m}^2$$

$$\Delta E = \frac{h^2}{4\pi^2 I} = 8.16 \times 10^{-22}\text{ J}$$

16-30 (a) Interval between lines is $0.624 \times 10^{12}\text{ s}^{-1} = \dfrac{h}{4\pi^2 I}$

$$I = \frac{h}{4\pi^2(0.624 \times 10^{12}\text{ s}^{-1})} = \frac{6.626 \times 10^{-34}\text{ J s}}{4\pi^2(0.624 \times 10^{12}\text{ s}^{-1})} = 2.69 \times 10^{-47}\text{ kg m}^2$$

(b) $0 - J$ transition: $\nu = \dfrac{h}{8\pi^2 I}[J(J+1) - 0] = \dfrac{h}{4\pi^2 I}\dfrac{J(J+1)}{2}$

$0 - 1$: $\nu = 0.624 \times 10^{12}\text{ s}^{-1}$; $\Delta E = h\nu = 4.13 \times 10^{-22}\text{ J}$

$0 - 2$: $\nu = 3 \times 0.624 \times 10^{12}\text{ s}^{-1}$; $\Delta E = h\nu = 1.24 \times 10^{-21}\text{ J}$

$0 - 3$: $\nu = 6 \times 0.624 \times 10^{12}\text{ s}^{-1}$; $\Delta E = h\nu = 2.48 \times 10^{-21}\text{ J}$

(c) Using isotope masses from Table 14.1 gives the reduced mass μ.

$$\mu = \frac{(1.0078\text{ g mol}^{-1})(34.969\text{ g mol}^{-1})(10^{-3}\text{ kg g}^{-1})}{(1.0078 + 34.969\text{ g mol}^{-1})(6.022 \times 10^{23}\text{ mol}^{-1})} = 1.627 \times 10^{-27}\text{ kg}$$

$$R_e = (I/\mu)^{1/2} = 1.29 \times 10^{-10}\text{ m} = 1.29\text{ Å}$$

(d) The absorptions are the 3-4, 4-5, 5-6, and 6-7 transitions.

16-32

$$\nu = \frac{c}{\lambda} = \frac{2.998 \times 10^8\text{ m s}^{-1}}{6.28 \times 10^{-5}\text{ m}} = 4.77 \times 10^{12}\text{ s}^{-1}$$

$$k/\mu = (2\pi\nu)^2$$

$$\mu = \frac{(22.9898\text{ g mol}^{-1})(22.9898\text{ g mol}^{-1})(10^{-3}\text{ kg g}^{-1})}{(2 \times 22.9898\text{ g mol}^{-1})(6.022 \times 10^{23}\text{ mol}^{-1})} = 1.91 \times 10^{-26}\text{ kg}$$

$$k = \mu(2\pi\nu)^2 = 17.2\text{ kg s}^{-2}$$

This is smaller than that for Li_2. The bond in Na_2 is weaker because the atoms are larger and thus farther apart.

16-34

$$\nu = \frac{c}{\lambda} = \frac{2.998 \times 10^8 \text{ m s}^{-1}}{2.9 \times 10^{-6} \text{ m}} = 1.03 \times 10^{14} \text{ s}^{-1}$$

$$k = \mu(2\pi\nu)^2 = (1.008 \text{ u})(1.66054 \times 10^{-27} \text{ kg u}^{-1})(2\pi \times 1.03 \times 10^{14} \text{ s}^{-1})^2$$

$$= 7.1 \times 10^2 \text{ kg s}^{-2}$$

This is almost 40% greater than the C–H force constant; the N–H bond is stiffer.

16-36 The energy difference is

$$\Delta E = h\nu = \left(6.626 \times 10^{-34} \text{ J s}\right)\left(1.15 \times 10^{13} \text{ s}^{-1}\right) = 7.62 \times 10^{-21} \text{ J}$$

$$\frac{P_i}{P_j} = \exp\left[\frac{-\Delta E}{k_B T}\right] = e^{-1.84} = 0.159$$

For every 1.00×10^6 molecules in the ground state, there are $0.159 \times 10^6 = 1.59 \times 10^5$ molecules in the first excited state.

16-38 The electron lost in the ionization of C_2H_4 will be from a bonding π-orbital because the π system in C_2H_4 is less than half filled, a fact that assures that all of the π-electrons are bonding electrons (in the ground state). The bond order between the two carbons in $C_2H_4^+$ is 3/2, less than that in C_2H_4.

16-40 Crystal violet is violet, which implies that it absorbs the color that is the complement of violet, namely yellow. This means it has a maximum in its absorption near 590 nm.

16-42

$$\Delta E = \Delta H - RT\Delta n_g = \Delta H - RT$$

$$= 211 \times 10^3 \text{ J mol}^{-1} - \left(8.315 \text{ J mol}^{-1} \text{ K}^{-1}\right)(298 \text{ K})$$

$$= 2.09 \times 10^5 \text{ J mol}^{-1} = 3.46 \times 10^{-19} \text{ J}$$

$$\Delta E = \frac{hc}{\lambda}$$

$$\lambda = \frac{hc}{\Delta E} = \frac{(6.626 \times 10^{-34} \text{ J s})(2.998 \times 10^8 \text{ m s}^{-1})}{3.46 \times 10^{-19} \text{ J}}$$

$$= 5.74 \times 10^{-7} \text{ m} = 574 \text{ nm}$$

16-44 (a) $C_{12}B_{24}N_{24}$ has the same number of electrons as C_{60}.

(b) Alternate double bonds in each hexagon are double bonds. There are two such resonant structures.

16-46 The energy 330 kJ mol^{-1} of the bonds is equivalent to 5.48×10^{-19} J per bond. Using $E = hc/\lambda$ gives $\lambda = 3.62 \times 10^{-7}$ m. This is 362 nm.

16-48 Carbon dioxide and sulfur dioxide have similar formulas, but quite different molecular structures. Carbon dioxide has a linear molecule. In it, the central C forms σ-bonds to the two oxygen atoms using sp hybrid orbitals. These bonds are joined by π bonds from the overlap of C-atom $2p$ and O-atom $2p$ orbitals. The central atom in CO_2 has no lone pairs. The molecule of SO_2 is a bent molecule. The central S is sp^2 hybridized: two of these orbitals overlap with orbitals of the O's in σ bonds and the third accommodates a lone pair. Thus, the SN of the central S is 3. The molecule is bent with an O—S—O angle near 120°. One $3p$ orbital of the S mixes with one $2p$ orbital from each of the O's in a π-system.

16-50 The carbide ion has 10 valence electrons. Its ground-state configuration is $(\sigma_{2s})^2(\sigma_{2s}^*)^2(\pi_{2p})^4(\sigma_{2p_z})^2$, the same as the configuration of N_2, and its bond order is 3.

16-52 The ground-state electron configuration of Be_2 is $(\sigma_{2s})^2(\sigma_{2s}^*)^2$. The predicted bond order of the molecule is zero, and according to simple theory it should not exist. The observed bond length and dissociation enthalpy of Be_2 are respectively high and quite low: the molecule indeed almost does not exist.

16-54 The electron configuration is $(\sigma_{1s})^2$ with the $1s$ orbital from He lying lower in energy than that from H. C_2 is larger than C_1 because the electron is more strongly attracted to the doubly positive He^{2+} nucleus.

16-56 The cyclic H_3^+ ion is a perfect equilateral triangle. It has two electrons. The electrons are in a low-energy bonding orbital mainly localized in the center of the triangle so as best to nullify internuclear repulsions.

16-58 *Trans*-tetrazene and *cis*-tetrazene have the structures:

The SN's of all nitrogen atoms in both structure are 3. There is sp^2 hybridization at all nitrogen atoms, and the molecules are planar, except for the terminal hydrogen atoms, because of the π system.

16-60

$$\mu = \frac{(22.9898 \text{ g mol}^{-1})(34.9689 \text{ g mol}^{-1})(10^{-3} \text{ kg g}^{-1})}{(22.9898 + 34.9689 \text{ g mol}^{-1})(6.022 \times 10^{23} \text{ mol}^{-1})} = 2.30334 \times 10^{-26} \text{ kg}$$

$$\nu = \frac{h}{8\pi^2 \mu R_e^2}[2(3) - 1(2)] = \frac{h}{2\pi^2 \mu R_e^2}$$

$$R_e^2 = \frac{h}{2\pi^2 \mu \nu} = \frac{6.626076 \times 10^{-34} \text{ J s}}{2\pi^2(2.30334 \times 10^{-26} \text{ kg})\nu}$$

For the ground state this gives $R_e = 2.36522$ Å, and for the excited state, $R_e = 2.37405$ Å

16-62 The ratio of probabilities is

$$\frac{P_i}{P_j} = \exp\left[\frac{-\Delta E}{k_B T}\right]$$

Taking $\Delta E = 2 \times 10^{-5}$ kJ mol^{-1} = 3.32×10^{-26} J gives

$$\frac{P_i}{P_j} = \exp\left(-8.07 \times 10^{-6}\right) = 1.00$$

The populations are almost equal. The same result holds if

$$\Delta E = 2 \times 10^{-4} \text{ kJ mol}^{-1}$$

16-64 The electron in ethylene is excited from a π to a π^* orbital. The transfer reduces the overall bond order of the the molecule by 1. The C to C bond in the molecule in the excited state should be *longer* than it was in the ground state. Because the bond is weaker, its force constant, k, is diminished, and the vibrational frequency of the C to C stretching mode is reduced.

16-66 (a) Excitation of the electron in H_2^+ to an antibonding orbital will lead to the dissociation of the molecule. The effect of an antibonding electron is to oppose the continued association of the nuclei.

(b) If even more energy is fed into a H_2^+ molecule, the electron can be brought into a higher-level bonding molecular orbital. The molecule should persist for a while in this bound state, but it will also eventually dissociate to H and H^+.

16-68 $S + O_2 \rightarrow SO_2$

$SO_2 + \frac{1}{2} O_2 \rightarrow SO_3$

$SO_3 + H_2O \rightarrow H_2SO_4$

$N_2 + O_2 \rightarrow 2\,NO$

$NO + \frac{1}{2} O_2 \rightarrow NO_2$

$NO_2 + OH \rightarrow HNO_3$

16-70 The greenhouse effect is the potential increase in the average temperature of the earth's surface by the re-radiation of infrared energy to the earth by molecules of such gases as methane and carbon dioxide in the atmosphere. Examples of CO_2-producing sources are coal and oil-burning power plants, automobile emissions and natural gas for home heating. Energy sources that do not produce such gases are nuclear power plants, solar energy sources, and hydroelectric plants.

16-72

$$C_{60}(s) + 60\,O_2(g) \rightarrow 60\,CO_2(g)$$

$$-25,891 \text{ kJ mol}^{-1} = 60(-393.51) - \Delta H_f^\circ(C_{60})$$

$$\Delta H_f^\circ(C_{60}) = +2280 \text{ kJ mol}^{-1}$$

16-74 The concentrations are

$$[CO] = \frac{P_{CO}}{RT} = 4.0 \times 10^{-11} \text{ mol L}^{-1}$$

$$[CH_4] = \frac{P_{CH_4}}{RT} = 6.8 \times 10^{-10} \text{ mol L}^{-1}$$

$$\frac{d[OH]}{dt} = -k_{CO}[CO][OH] - k_{CH_4}[CH_4][OH]$$

$$= -(1.6 \times 10^{11} \text{ L mol}^{-1} \text{ s}^{-1})(4.0 \times 10^{-11} \text{ mol L}^{-1})[OH]$$

$$-(3.8 \times 10^9 \text{ L mol}^{-1} \text{ s}^{-1})(6.8 \times 10^{-10} \text{ mol L}^{-1})[OH]$$

$$= -9.0 \text{ s}^{-1}[OH] = -k_{\text{eff}}[OH]$$

The half life is

$$t_{\frac{1}{2}} = \frac{\ln 2}{k_{\text{eff}}} = 0.077 \text{ s}$$

Chapter 17

Bonding, Structure, and Reactions of Organic Molecules

17-2 The structure of acrylonitrile is:

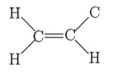

The molecule is planar with the six bond angles at the two double-bonded car-bon atoms (both sp^2-hybridized) all close to 120° and the bond angle at the triple-bonded C atom (sp-hybridized) equal to 180°. The p-orbitals perpendicu-lar to the molecular plane form a π-bonding system with 4 electrons resembling that in butadiene (Fig. 17.5). The other two p-orbitals on the sp-hybridized C atom and the N atom form an additional doubly occupied π-orbital.

17-4 (a)

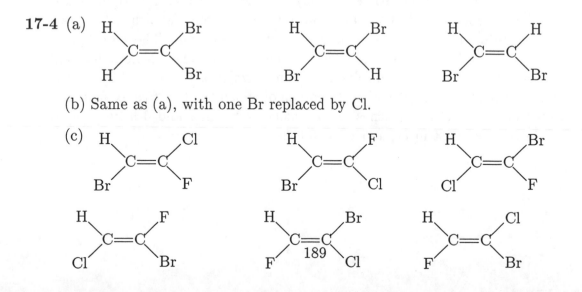

(b) Same as (a), with one Br replaced by Cl.

(c)

17-6 (a) The standard enthalpy of formation ΔH_f° of C_4H_8 is the enthalpy change in forming C_4H_8 in its standard state from $C(s)$ and $H_2(g)$ in their standard states: $4\,C(s) + 4\,H_2(g) \rightarrow C_4H_8(g)$. Imagine this reaction to proceed by the atomization of 4 mol of $C(s)$ and 4 mol of $H_2(g)$ to $C(g)$ and $H(g)$ followed by the formation of 4 mol of C—C bonds and 8 mol of C—H bonds. The sum of the ΔH°'s of these steps equals the desired ΔH°, by Hess's law.

$$\Delta H^{\circ} = 4(716.7) + 8(218.0) + 4(-348) + 8(-413) = -85 \text{ kJ}$$

The estimated ΔH_f° of cyclobutane based on bond enthalpies is -85 kJ mol^{-1}.

(b) The standard enthalpy change for the combustion of cyclobutane is the enthalpy of formation of the products minus the enthalpy of formation of the reactants:

$$\Delta H^{\circ} = -2568 \text{ kJ} = 4(-393.51) + 4(-241.82) - 6(0) - x$$

where x is the exact ΔH_f° of cyclobutane. Solving gives $\Delta H_f^{\circ} = +27$ kJ mol^{-1}.

(c) The actual ΔH_f° (part b) is greater than the estimated ΔH_f° (part a). It follows that making the ring of four carbon atoms is endothermic by about 112 kJ mol^{-1}. This is the "strain enthalpy."

17-8 (a) The catalytic cracking reaction is:

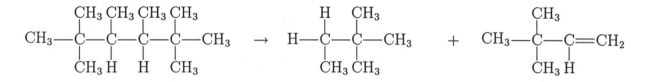

(b) There are 16 alkene isomers. They include three butenes, eight pentenes, and five hexenes. The butenes are 3,3-dimethyl-1-butene (shown in the previous part) 2,3-dimethyl-1-butene, and 2,3-dimethyl-2-butene. The eight pentenes include three 1-pentenes, which are 2-methyl-1-pentene, 3-methyl-1-pentene and 4-methyl-1-pentene, and five 2-pentenes, which are 2-methyl-2-pentene, 3-methyl-2-pentene and 4-methyl-2-pentene of which the last two exist in *cis* and *trans* forms. The five hexenes are 1-hexene, 2-hexene, and 3-hexene of which the latter two exist in *cis* and *trans* forms

17-10 The structural formulas are:

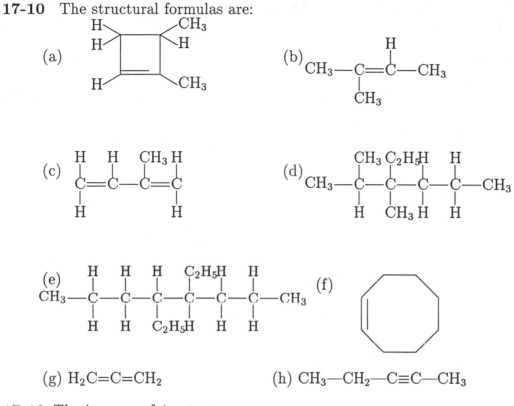

(g) $H_2C=C=CH_2$ (h) $CH_3-CH_2-C\equiv C-CH_3$

17-12 The isomers of 4-octene are:

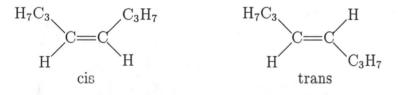

cis   trans

17-14 (a) 2,3-dimethyl-1,3-butadiene (b) 2,4-hexadiene (c) 2,2-dimethylbutane
(d) methylpropene

17-16 (a) The two -CH_3 carbons are sp^3, and the other four are sp^2 hybridized.

(b) The two -CH_3 carbons are sp^3, and the other four are sp^2 hybridized.

(c) All are sp^3 hybridized.

(d) The two -CH_3, carbons are sp^3, and the other two are sp^2 hybridized.

17-18 (a)

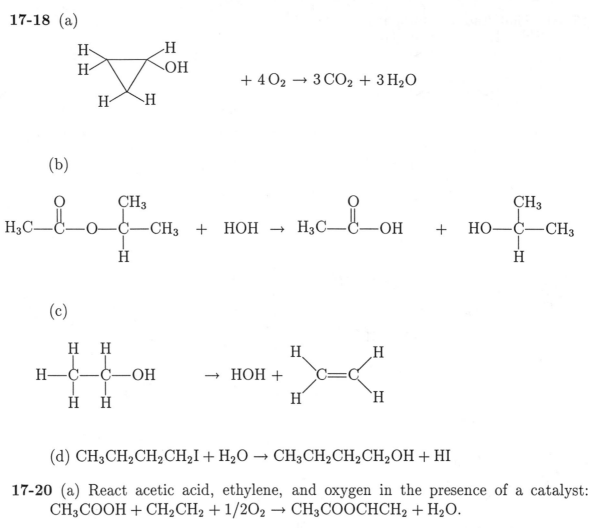

$$+ 4\,O_2 \rightarrow 3\,CO_2 + 3\,H_2O$$

(b)

(c)

(d) $CH_3CH_2CH_2CH_2I + H_2O \rightarrow CH_3CH_2CH_2CH_2OH + HI$

17-20 (a) React acetic acid, ethylene, and oxygen in the presence of a catalyst: $CH_3COOH + CH_2CH_2 + 1/2O_2 \rightarrow CH_3COOCHCH_2 + H_2O$.

(b) React formic acid and ammonia: $HCOOH + NH_3 \rightarrow HCO(NH_2) + H_2O$.

(c) React ethylene with fluorine: $H_2CCH_2 + F_2 \rightarrow FCH_2CH_2F$.

17-22 A tertiary amine has no hydrogen atoms on the nitrogen atom. Hence, it cannot contribute hydrogen to a water molecule as do primary and secondary amines when they condense with a carboxylic acid.

17-24 Compute the amount of ethylene dichloride required to make the 3.73×10^9 kg of vinyl chloride. It is clear without writing an equation that the molar ratio of the vinyl chloride to the ethylene dichloride is 1 to 1. Also, 1 mol of HCl is produced for every 1 mol of vinyl chloride. Then:

$$3.73 \times 10^9 \text{ kg } C_2H_3Cl \times \left(\frac{1 \text{ mol } C_2H_3Cl}{0.06250 \text{ kg } C_2H_3Cl} \right) \times \left(\frac{1 \text{ mol } C_2H_4Cl_2}{1 \text{ mol } C_2H_3Cl} \right)$$

$$\times \left(\frac{0.09896 \text{ kg } C_2H_4Cl_2}{1 \text{ mol } C_2H_4Cl_2} \right) = 5.91 \times 10^9 \text{ kg } C_2H_4Cl_2$$

This is 94.3% of the total production of ethylene dichloride. The mass of the by-product HCl is 2.18×10^9 kg, by a similar computation.

17-26 (a) Paramagnetic

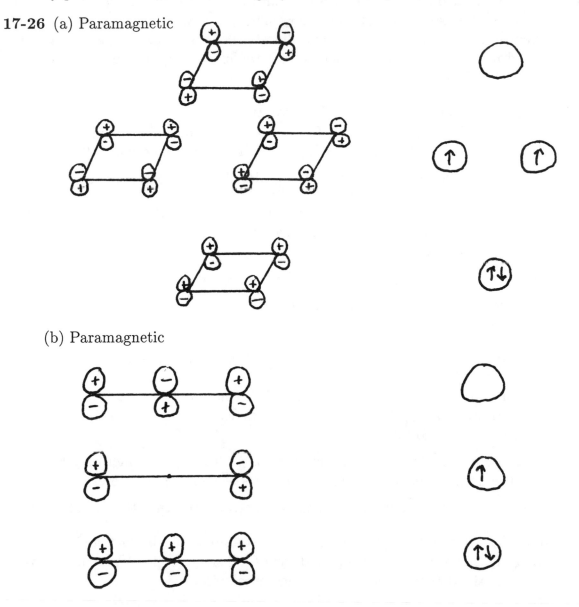

(b) Paramagnetic

17-28 (a) There are five candidate structures of benzene. Structures (i), (ii), and (iv) would form only one C_6H_5Cl. The other two proposed structures accommodate Cl in place of H in at least two distinct locations.

(b) Only structure (ii) gives exactly three isomers of formula $C_6H_4Cl_2$. Structures (i), (iii), and (v) give more than three isomers, and structure (iv) gives only two isomers.

17-30 Make an alcohol that is enriched in an isotope of oxygen, such as $R(^{18}O)H$. Prepare the ester and determine how much, if any, of the heavier isotope is incorporated in it. It is found that almost all of the labeled oxygen is incorporated in the ester, and we conclude that the oxygen in an ester comes entirely from the alcohol rather than the carboxylic acid.

17-32 (a) The two equations are both dehydrations:
$CH_3CH_2OH \rightarrow H_2C{=}CH_2 + H_2O$ over alumina, 400 °C
$CH_3CH_2OH \rightarrow CH_3CH_2{-}O{-}CH_2CH_3 + H_2O$ over alumina, 230°C

(b) The reaction is a dehydrogenation: $CH_3CH_2OH \rightarrow CH_3CH{=}O + H_2$.

17-34 (a) The critical temperature of ethylene is 9.50°C, which is 49.10°F. The critical pressure is 736.21 pounds per square inch.

(b) At 54 atm and 15°C, the ethylene in the pipeline is a supercritical fluid—both the critical temperature and critical pressure are exceeded.

(c) The volume of the pipeline is its interior cross-sectional area times its length. Expressing all dimensions in meters gives:

$$V = \pi r^2 l = 3.14159(125 \times 10^{-3} \text{ m})^2(1.00 \times 10^5 \text{ m}) = 4.91 \times 10^3 \text{ m}^3$$

This is 4.91×10^6 L. At 54 atm and 15°C it contains 1.12×10^7 mol of an ideal gas. Multiplying by the molar mass of ethylene ($\mathcal{M} = 28.054$ g mol^{-1}) gives 3.14×10^8 g, or 314,000 kg.

(d) The actual density of the ethylene is 0.20 g cm^{-3}, which is 200 g L^{-1}. Multiplying by the volume of the pipe gives 982,000 kg, 3.1 times greater.

(e) The operator should increase the pressure to keep the ethylene all liquid.

17-36 For rotation to take place about the central bond in 2-butene, a C=C double bond must be broken and replaced by a C–C single bond. The energy cost is close to the difference of the two bond enthalpies:

$$\Delta E = 615 - 348 = 267 \text{ kJ mol}^{-1}$$

This is much higher than the energy barrier given in Section 16.3 for the inversion of ammonia, 24 kJ mol^{-1}, so the isomerization of 2-butene will be much slower than the inversion of ammonia.

17-38 The more widely delocalized π-electrons in naphthalene have longer characteristic wavelengths than the π-electrons in benzene. The wavelength of maximum absorption of light will accordingly be shifted to a wavelength longer than 255 nm (see text Table 16.4).

Chapter 18

Bonding in Transition Metals and Coordination Complexes

18-2 $TiCl_4$ and $TiBr_4$ are both covalent compounds. The latter has stronger inter-molecular forces (and higher melting point) because it has more electrons to provide long-range attractions. On the other hand, TiF_4 has a higher melting point because it has more ionic character than $TiCl_4$.

18-4 The chemical formula is V_2O_3, and this should be more basic than the higher oxide V_2O_5 from problem 18-3.

$$V_2O_3(s) + 6\ H_3O^+(aq) \rightarrow 2\ V^{3+}(aq) + 9\ H_2O(l)$$

18-6

$$6\ CoO + O_2 \rightarrow 2\ Co_3O_4$$

The average oxidation state of Co in Co_3O_4 is $\frac{8}{3}$. This corresponds to two Co^{3+} and one Co^{2+} ion.

18-8 The glycinate ion ($H_2N-CH_2-COO^-$) has two donor sites. They are the nitrogen, which has one lone pair of electrons, and the carboxylate oxygen, which has three lone pairs. The ligand can bind to a metal ion by donating electron pairs from either of these sites or both.

18-10 The oxidation state of Mn in $Mn_2(CO)_{10}$ is zero; the oxidation state of Re in $[Re_3Br_{12}]^{3-}$ is +3; the oxidation state of the Fe in $[Fe(H_2O)_4(OH)_2]^+$ is +3; the oxidation state of the Co in $[Co(NH_3)_4Cl_2]^+$ is + 3.

18-12 (a) $Ag_4[Fe(CN)_6]$ (b) $K_2[Co(NCS)_4]$ (c) $Na_3[VF_6]$ (d) $K_3[Cr(C_2O_4)_3]$

18-14 (a) tetraaquadihydroxonickel(II)
(b) chloroiodomercury(II)
(c) potassium hexacyanoosmate(II)
(d) bromochlorobis(ethylenediamine)iron(III) chloride

18-16 Such a complex suffers ligand exchange at only a very slow rate (the meaning of "inert"). It however does tend to decompose spontaneously in some fashion (the meaning of "thermodynamically unstable").

18-18 We combine ΔG_f°'s in the usual way to determine the ΔG° of the reaction as it is given in the problem:

$$\underbrace{1(-45.6)}_{1\,\text{Ni}^{2+}(aq)} + \underbrace{4(172.4)}_{4\,\text{CN}^-(aq)} - \underbrace{1(+472.1)}_{1\,[\text{NiCN}]_4^{2-}(aq)} = 171.9 \text{ kJ}$$

A positive ΔG° means that the complex is thermodynamically stable with respect to decomposition to aquated metal ion and ligands, although exchange of the ligands occurs rapidly. The equilibrium constant is found from:

$$\ln K = \frac{-\Delta G^\circ}{RT} = \frac{-171.9 \times 10^3 \text{ J mol}^{-1}}{(8.315 \text{ J mol}^{-1}\text{ K}^{-1})(298.15 \text{ K})} = -69.34$$

The K is 7.7×10^{-31}.

18-20 Conductivity increases in the order

$$[\text{Fe(NH}_3)_3\text{Cl}_3] < [\text{Cr(NH}_3)_4\text{Cl}_2]\text{Cl} < \text{BaCl}_2 < \text{K}_4[\text{Fe(CN)}_6]$$

18-22 (a) The square-planar complex bromochloro(ethylenediamine)platinum(II) has only one isomer because the $\text{NH}_2\text{CH}_2\text{CH}_2\text{NH}_2$ ligand is not long enough to span *trans* positions. The structure is:

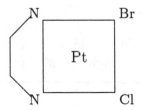

(b) The complex ion has *cis* and *trans* isomers just like those of text Figure 18-9 except with a central iron(III) ion.

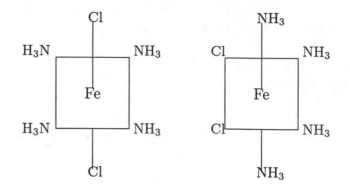

(c) The amminechlorobis(ethylenediamine)iron(III) ion, a +1 ion, has two *cis* forms and one *trans* form. Both of the *cis* forms (center and right in the following) are chiral, that is, non-superimposable upon their mirror images. This makes a total of five isomers, of which three are:

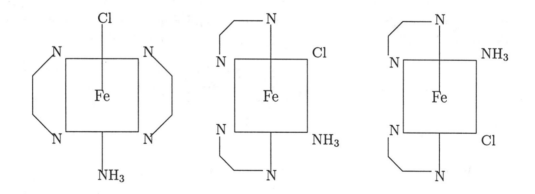

18-24 There are two each of the three ligands NH_3, Cl^-, and F^-. The members of each pair may be either *cis* or *trans* about the central Pt(IV) in octahedral coordination. This would seem to mean that there are eight isomers because there are eight possible triple combinations of *cis* and *trans* (*cis-cis-cis, cis-cis-trans, cis-trans-cis,* and so forth). But, placing two pairs *trans* forces the third pair to be *trans*, also. This eliminates *cis-trans-trans, trans-cis-trans, and cis-trans-trans* from the list of eight. Only five *cis-trans* isomers are therefore possible:

NH_3	*cis*	*cis*	*cis*	*trans*	*trans*
Cl	*cis*	*cis*	*trans*	*cis*	*trans*
F	*cis*	*trans*	*cis*	*cis*	*trans*

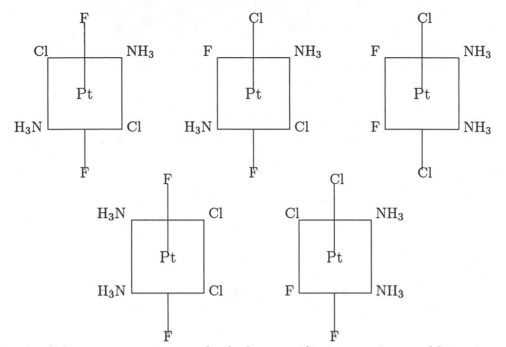

Each of these isomers is now checked to see if it is superimposable on its mirror image. Any isomer that has even one *trans* disposition of ligands has a plane of symmetry and *is* superimposable on its mirror image. The applies to four of the above five cases. The *cis-cis-cis* geometry (bottom right) is *not* superimposable on its mirror image and exists as a pair of enantiomers.

18-26 The splitting diagram to use in every case the right-hand part of Figure 18.17.

(a) The Cr^{2+} ion is a d^4 case. In a strong octahedral field it has the ground-state configuration t_{2g}^4 for two unpaired electrons; in a weak octahedral field the configuration is $t_{2g}^3 e_g^1$ and all four electrons are unpaired.

(b) The V^{3+} ion is a d^2 case. It has two unpaired electrons (electron configuration t_{2g}^2) in both a strong and a weak octahedral field.

(c) The Ni^{2+} ion is a d^8 case. It has the configuration $t_{2g}^6 e_g^2$ for two unpaired electrons in both a strong and a weak octahedral field.

(d) The Pt^{4+} ion is a d^6 case. It has four unpaired electrons in a weak octahedral field ($t_{2g}^4 e_g^2$) and no unpaired electrons (t_{2g}^6) in a strong octahedral field.

(e) The Co^{2+} ion is a d^7 case. It has three unpaired electrons in a weak octahedral field ($t_{2g}^5 e_g^2$) and one unpaired electron ($t_{2g}^6 e_g^1$) in a strong octahedral field.

18-28 The Mn(III) ion in these two complexes has four *d*-electrons. When coordinated by Cl^- in the $[MnCl_6]^{3-}$ ion, the result is a high-spin complex–all four electrons are unpaired. With a stronger-field ligand in the $[Mn(CN)_6]^{3-}$, there are no electrons in the higher-energy e_g-levels. Putting four electrons into the t_{2g} level requires pairing two. The other two are unpaired. In the first case, the CFSE is -3/5Δ_o; in the second it is -8/5Δ_o.

18-30 The problem concerns the standard reduction potentials of $Mn^{3+}(aq)$, $Fe^{3+}(aq)$ and $Co^{3+}(aq)$ to the corresponding 2+ ions. All six of the ions are octahedrally coordinated by H_2O molecules in acidic aqueous solution. The metals in the 3+ ions are d^4, d^5, and d^6 species; the metals in the 2+ ions are d^5, d^6 and d^7 species, respectively. These hexaaqua complexes are all high-spin complexes because water is a weak-field ligand. The reduction potential of $Fe^{3+}(aq)$ is less than the reduction potential of its neighbors in the periodic table, which means that it is *harder* to reduce to the +2 state than the neighbors. The $Fe^{3+}(aq)$ ion has the high-spin $t_{2g}^3 e_g^2$ configuration. This is a relatively stable electron configuration relative to configurations with four or six d-electrons (note that both the e_g and the t_{2g} levels of Fe^{3+} are half-filled). As a consequence, the $Fe^{3+}(aq)$ ion resists reduction better.

18-32 The ion $[PtI_6]^{2-}$ absorbs all across the visible spectrum.

18-34 A solution of hexaaquanickel(II) ion is green. It therefore absorbs the complementary color of green, which is red. We can estimate the wavelength of the absorption in the red as 700 nm (see text Figure 15.3). The corresponding energy is

$$E = \frac{hc}{\lambda} = \frac{(6.626 \times 10^{-34} \text{ J s})(3.00 \times 10^8 \text{ m s}^{-1})}{700 \times 10^{-9} \text{ m}} = 2.8 \times 10^{-19} \text{ J}$$

The absorption occurs with the excitation of an electron from a t_{2g} to an e_g level so this energy is Δ_o. Multiplying Δ_o by Avogadro's number puts it on a molar basis. It is about 170 kJ mol^{-1}

18-36 The CFSE for this d^8 complex is $-\frac{6}{5}\Delta_0$, or about -200 kJ mol^{-1}.

18-38 (a) The complement of yellow is violet; the hexaamminecobalt(III) ion therefore absorbs in the violet.

(b) We can estimate the wavelength from the wavelength of the complement of the observed color. It is probably near 410 nm as a maximum (see text Figure 15.3).

18-40 (a) The complex $[Mn(H_2O)_6]^{2+}$ results from the dissolution of $Mn(NO_3)_2$ in water. It is a high-spin d^5 complex. The excitation of an electron requires the reversal of a spin; such processes are fairly strictly forbidden. Their rarity makes the color of the ion faint. The $[Mn(CN)_6]^{4-}$ ion on the other hand is a low-spin complex. The $t_{2g} \to e_g$ transitions are not spin-forbidden and are more intense.

(b) We expect $Zn(NO_3)_2$, $CdSO_4$ and $AgClO_3$ to be colorless in aqueous solution based on the full occupancy of the d-orbitals of the metal ion when coordinated by the solvent.

18-42 All of the complexes in the problem follow the rule of 18 with the exception of $[V(CO)_6]$, which has 17 valence electrons surrounding the central vanadium.

18-44 The mass of iron in one mole of hemoglobin is 0.33% of the mass of one mole of hemoglobin, which is given as 6.8×10^4 g. This comes out to 224 g of iron. Dividing this mass of iron by the molar mass of iron establishes that there are four moles of iron in one mole of hemoglobin. Hence there are four atoms of iron per hemoglobin molecule.

18-46 Let M stand for the metal (Ni, Cu, or Zn). Then we can use the following four steps (with data from Table 18-1) to represent the overall process taking solid metal to solution:

		Ni	Cu	Zn
$M(s) \to M(g)$	ΔH	430	338	131
$M(g) \to M^+(g) + e^-$	ΔE	737	745	906
$M^+(g) \to M^{2+}(g) + e^-$	ΔE	1753	1958	1733
$M^{2+}(g) \to M^{2+}(aq)$	ΔH	-2985	-2989	-2937
$M(s) \to M^{2+}(aq) + 2\ e^-$		-65	52	-167

In this table we have mixed energy and enthalpy changes, so the sums in the bottom line are not absolute energy or enthalpy changes. They are useful only for comparison purposes, and show the stability of $Cu(s)$ relative to $Ni(s)$ and $Zn(s)$, or, equivalently, the greater tendency of $Cu^{2+}(aq)$ to be reduced to $Cu(s)$.

Looking at the table itself, the number that stands out as being primarily responsible for the positive +52 for Cu is the second ionization energy IE_2. This is much higher for Cu than for Ni or Zn.

18-48 The reaction is:

$$2\,[Cr(en)_2(NCS)_2]SCN(s) \to [Cr(en)_2(NCS)_2]\,[Cr(en)(NCS)_4](s) + en(g)$$

The oxidation number of the chromium does not change from +3 in this reaction. Rather, two NCS^- anions replace an ethylenediamine in the coordination sphere of one of the complexes.

18-50 $2(NH_4)_2[Ir(H_2O)Cl_5] \rightarrow NH_3(g) + 2H_2O(g) + HCl(g) + (NH_4)_3[Ir_2Cl_9]$

ammonium aquapentachloroiridate(III).

18-52 (a) $[Fe(H_2O)_4Cl_2]CO_3$ is most likely to have the same electrical conductivity per mole as $MgSO_4$, which is also a 2 to 2 ionic compound.
(b) $[Mn(H_2O)_6]Cl_3$ matches best to $GaCl_3$, which also has Cl^- as the anion, but also matches to Na_3PO_4.
(c) $[Zn(H_2O)_3(OH)]Cl$ matches best to $NaCl$.
(d) $[Fe(NH_3)_6]_2(CO_3)_3$ matches best to $Fe_2(SO_4)_3$.
(e) $[Cr(NH_3)_3Br_3]$ matches best to HCN; both are nonelectrolytes.
(f) $K_3[Fe(CN)_6]$ matches well to both Na_3PO_4, as well as to $GaCl_3$.

18-54 A planar hexagonal structure for $[Co(NH_3)_4Cl_2]^+$ would have three isomers: one with the two Cl's next to each other, one with the two Cl's separated by one NH_3, and one with the two Cl's separated by two NH_3's around the hexagonal ring. A trigonal prismatic structure would have three isomers as well: one with both Cl's on the same triangular end of the prism, one with the two Cl's on opposite triangular ends and lying on the same edge, and one with the two Cl's on opposite ends and lying on different edges. The last of these would be chiral.

18-56 The compound has an odd number of valence electrons and will be paramagnetic.

18-58 In a crystal field of octahedral symmetry all three p orbitals have the same energy. There is no splitting. Electrons in the p-orbitals have their greatest probability density along a coordinate axis and differ among themselves only in which axis. The axes are indistinguishable in the octahedral case; each has one ligand on each end. A square-planar field is derived from an octahedral field by removal of ligands from one axis, say, the z. If so, the p_z orbital would be split to lower energy (since it would not encounter ligands directly) while the p_y and p_x orbitals would remain unsplit.

18-60 Mn^{2+} is not a strong reducing agent because it appears high up on the right side in Appendix E (it is difficult to <u>oxidize</u> Mn^{2+} to Mn^{3+} because the <u>reduction</u> potential of the reverse reaction, 1.51 V, is so high).

Mn^{2+} is not a strong oxidizing agent because it appears far down on the left side of Appendix E (it is difficult to reduce it because its reduction potential to $Mn(s)$, -1.029 V, is so negative).

We conclude that $Mn^{2+}(aq)$ is unusually stable compared to other metal ions.

18-62 The two compounds have the same number of unpaired electrons and both have Mn^{2+}, a d^5-metal ion, as the central metal ion. The octahedral complex is a weak-field, high-spin complex and so is the tetrahedral complex.

18-64 (a) The compound $(NH_4)_2[Fe(H_2O)F_5]$ has an odd number of valence electrons and is paramagnetic regardless on the strength of the ligands. Because the ligands are weak-field ligands, the octahedral field is weak and the five d-electrons of the Fe(III) should remain unpaired. This makes the compound paramagnetic to the extent of five unpaired electrons.

(b) The anion is a high-spin d^5 complex like $[Fe(H_2O)_6]^{3+}$ so the complex is probably a pale violet. Note that the cation (the ammonium ion) is colorless.

(c) The electron configuration of the d-electrons of the iron is $t_{2g}^3 e_g^2$.

(d) The compound is named ammonium aquapentafluoroferrate(III).

18-66 The C$-$O bond length should be increased in $Ni(CO)_4$, relative to the free molecule, because of π-back-bonding. This has the effect of delocalizing the $3d^8$ electrons of Ni over the π^* orbital of each CO, thereby lowering its bond energy and increasing its bond length.

18-68 The compound $[V(N_2)_6]$ should be octahedral with the six ligands arranged around the central V atom. The compound has an odd number of valence electrons and is paramagnetic. It is isoelectronic with $[V(CO)_6]$.

18-70 Recall from Section 18.6 the stability of the $C_5H_5^-$ ion. After tungsten has donated two electrons to form a pair of such ions, it is a d^4 compound, while tantalum is d^3. After W shares two electrons with bonded H-atoms to form single bonds, it has two electrons left to form a single bond with an approaching H^+ ion; it is thus a good base. By contrast, Ta has already shared its three valence electrons with H-atoms and has little affinity for an additional H^+ ion.

18-72 The reaction is

$$2\ Sc(s) + 6\ H_3O^+(aq) \rightarrow 2\ Sc^{3+}(aq) + 6\ H_2O(l) + 3\ H_2(g)$$

$$\left(\frac{4.53\ g\ Sc}{44.956\ g\ mol^{-1}}\right) \times \left(\frac{3\ mol\ H_2}{2\ mol\ Sc}\right) = 0.151\ mol\ H_2(g)$$

Assuming 1 atm pressure and 25°C, the volume is

$$V = \frac{nRT}{P} = \frac{(0.151\ mol)(0.08206\ L\ atm\ mol^{-1}K^{-1})(298\ K)}{1.00\ atm} = 3.70\ L$$

18-74 The equilibrium is

$$2\,CrO_4^{2-}(aq) + 2\,H_3O^+(aq) \rightleftharpoons Cr_2O_7^{2-}(aq) + 3\,H_2O(l)$$

$$\frac{[Cr_2O_7^{2-}]}{[CrO_4^{2-}]^2[H_3O^+]^2} = K = 4.2 \times 10^{14}$$

Set $[H_3O^+] = 1.0 \times 10^{-7}$ (at pH 7.00) and take $[Cr_2O_7^{2-}] = 0.10 - x$, with $[CrO_4{}^{2-}] = 2x$. Then

$$\frac{0.10 - x}{(2x)^2(1.0 \times 10^{-7})^2} = 4.2 \times 10^{14}$$

$$\frac{0.10 - x}{x^2} = 16.8$$

Solving gives $x = 0.0529$

$$[CrO_4^{2-}] = 2x = 0.11\ M$$

18-76 The chemical amount of CO present is

$$n = \frac{PV}{RT} = \frac{(2.00\ atm)(1.18\ L)}{(0.08206\ L\ atm\ mol^{-1}K^{-1})(298\ K)} = 0.09646\ mol\ CO$$

The mass of CO in this is

$$(0.09646\ mol\ CO)(28.01\ g\ mol^{-1}) = 2.70\ g$$

so the mass of the remaining osmium is

$$6.79 - 2.70 = 4.09\ g\ Os$$

The chemical amount of Os is

$$\frac{4.09\ g\ Os}{190.2\ g\ mol^{-1}} = 0.02149\ mol\ Os$$

The ratio is

$$\frac{0.09646}{0.02149} = 4.49 \approx 4.5 = \frac{9}{2}$$

so the probable empirical formula is $Os_2(CO)_9$. This compound might have a single bridging CO ligand, with four other CO molecules attached separately to each Os atom.

Chapter 19

Structure and Bonding in Solids

19-2 (a) A cereal box does <u>not</u> have four-fold rotational symmetry since no two perpendicular edges are equal in length.

(b) A stop sign (octagonal in shape) does have a four-fold rotational axis of symmetry (perpendicular to the plane of the sign).

(c) A tetrahedron (solid with equivalent equilateral triangles at its sides) does <u>not</u> have a four-fold axis of symmetry; is does have four differend three-fold axes.

(d) A cube has three four-fold rotational axes (passing through the centers of its three pair of parallel faces).

19-4 The symmetry elements of the PF_5 molecule are one 3-fold axis of rotation (through the axial F atoms), three 2-fold axes of rotation (through the P atom and the equatorial F atoms), one horizontal plane of symmetry (reflection), and three vertical symmetry planes (reflection).

19-6 The problem gives 2θ, λ, and n ready for substitution in the (rearranged) Bragg law:

$$d = \frac{n\lambda}{2\sin\theta} = \frac{2(1.237\ \text{Å})}{2\sin(35.58/2)} = 4.049\ \text{Å}$$

19-8 Rearrange the Bragg law and substitute:

$$\sin\theta = \frac{n\lambda}{2d} = \frac{2(1.539\ \text{Å})}{2(4.28\ \text{Å})} = 0.3596$$

The inverse sine of 0.3596 is θ; $2\theta = 42.1°$.

19-10 Write the Bragg law as $\sin\theta = n\lambda/2d$ and substitute the given values of λ and d. This gives $\sin\theta = 0.07160n$. Diffracted beams are to be found at 2θ's corresponding to integral values of n. The answers are

n	$\sin\theta$	2θ	n	$\sin\theta$	2θ	n	$\sin\theta$	2θ
1	0.07160	8.21°	6	0.4296	50.88°	11	0.7876	103.9°
2	0.1432	16.47	7	0.5012	60.16	12	0.8592	118.4
3	0.2148	24.81	8	0.5728	69.89	13	0.9308	137.1
4	0.2864	33.28	9	0.6444	80.24	14	1.0024	–
5	0.3580	41.95	10	0.7160	91.45			

There are exactly 13 possible diffracted beams of this wavelength from this set of planes. If $n \geq 14$ then $\sin\theta > 1.00$, and θ is not defined.

19-12 The formula for the volume of a general parallelepiped is in the text. It is

$$V = abc\sqrt{1 - \cos^2\alpha - \cos^2\beta - \cos^2\gamma + 2\cos\alpha\cos\beta\cos\gamma}$$

The crystal is monoclinic, so $\alpha = \gamma = 90°$. The cosine of 90° is zero, hence, using the trigonometric identity $\sin^2\beta + \cos^2\beta = 1$:

$$V = abc\sqrt{1 - \cos^2\beta} = abc\sin\beta$$

Substitution gives $V = 589\ \text{Å}^3$.

19-14 The volume of the unit cell of strontium chloride hexahydrate is determined by substituting in the general formula for the volume of a unit cell with the understanding that in a triagonal cell $a = b = c$ and $\alpha = \beta = \gamma$. Substitution of the a and α give $V = 678.413\text{Å}^3 = 678.413 \times 10^{-24}\ \text{cm}^3$. Meanwhile, the molar mass of $SrCl_2 \cdot 6H_2O$ is 266.617 g mol^{-1}. A mole unit of this crystal therefore weighs $266.617 \times 3 = 799.851$ g because there are three formula units per cell. The volume of a mole of unit cells is the volume of a single cell multiplied by 6.022137×10^{23} (Avogadro's number) and equals 408.54 cm^3. The density of the substance is the mass of the mole of unit cells divided by its volume. It is 1.958 g cm^{-3}. The measured density of $SrCl_2 \cdot 6H_2O$ is 1.93 g cm^{-3}.

19-16 Suppose that we have 1 cm^3 of iron. It weighs 7.86 g. The volume of this iron can also be expressed in cubic ångstroms: 10^{24} Å^3. How many unit cells are in this volume? Each unit cell has a volume of $(2.87)^3$ Å$^3 = 23.64$ Å^3. It follows that there are 4.230×10^{22} unit cells in the 1 cm^3 of iron. There are two lattice points per unit cell, because the cells are body-centered. With an iron atom at every lattice point, there are 8.46×10^{22} atoms of iron in the specimen. Each Fe atom weighs 55.847 u (the atomic mass of Fe expressed in atomic units). The 1 cm³ of iron therefore has a mass of 4.7248×10^{24} u. But the ratio of the mass of a object in u to its mass in grams is Avogadro's number. Dividing 4.7248×10^{24} by 7.86 gives 6.01×10^{23} as Avogadro's number.

19-18 We calculate the volume of the unit cell of turquoise by substitution of $\alpha = 68.61°$, $\beta = 69.71°$, $\gamma = 65.08°$, $a = 7.424$ Å, $b = 7.629$ Å, and $c = 9.910$ Å into the formula:

$$V = abc\sqrt{1 - \cos^2\alpha - \cos^2\beta - \cos^2\gamma + 2\cos\alpha\cos\beta\cos\gamma} = 461.4 \text{ Å}$$

Now, suppose we have a (mammouth!) turquoise that contains one mole of unit cells. The volume of this turquoise is:

$$461.4 \text{ Å}^3 \times \left(\frac{1 \text{ cm}^3}{10^{24} \text{ Å}^3}\right) \times 6.022 \times 10^{23} = 277.86 \text{ cm}^3$$

This mass of the stone is its volume multiplied by its density (2.927 g cm^{-3}) or 813.3 g. Thus we have the mass of a mole of unit cells of turquoise. We also have the formula mass of turquoise by adding up the contributions of the several elements in the formula given in the problem. It is 813.44 g mol^{-1}. Since $813.3 \approx 813.44$, every unit cell contains one formula unit. There is one Cu atom per formula unit, so every unit cell contains one Cu atom.

19-20 A unit cell has eight corners, but each is shared by a total of eight cells, so it contributes $\frac{1}{8} \times 8 = 1$ Ca atom per unit cell. There is one Ti atom at each cell center. There are six faces, each shared by two unit cells, so there are $\frac{1}{2} \times 6 = 3$ oxygen atoms per unit cell. The chemical formula is $CaTiO_3$.

19-22 (a) 4 atoms per unit cell

(b)

$$V_m = \frac{26.98 \text{ g mol}^{-1}}{2.70 \text{ g cm}^{-3}} = \frac{N_0 a^3}{n}$$

$$n = 4 \text{ atoms per unit cell}$$

$$a^3 = \frac{4 \times 26.98 \text{ g mol}^{-1}}{6.022 \times 10^{23} \text{ mol}^{-1} \times 2.70 \text{ g cm}^{-3}} = 6.637 \times 10^{-23} \text{ cm}^3$$

$$a = 4.05 \times 10^{-8} \text{ cm}^3$$

$$d = \frac{a\sqrt{2}}{2} = 2.86 \times 10^{-8} \text{ cm}^3$$

19-24 (a)

$$V_{\rm m} = \frac{58.69 \text{ g mol}^{-1}}{8.90 \text{ g cm}^{-3}} = \frac{N_0 a^3}{4}$$

$$a^3 = \frac{4 \times 58.69 \text{ g mol}^{-1}}{8.90 \text{ g cm}^{-3} \times 6.022 \times 10^{23} \text{ mol}^{-1}} = 4.380 \times 10^{-23} \text{ cm}^3$$

$$a = 3.525 \times 10^{-8} \text{ cm}$$

$$d = \frac{a\sqrt{2}}{2} = 2.49 \times 10^{-8} \text{ cm}$$

(b) The radius of a Ni atom is $a\frac{\sqrt{2}}{4}$, which is 1.25×10^{-8} cm.

(c) The interstitial site in the f.c.c. unit cell lies with its center at the geometric center of the unit cell. Accordingly, the radius of the octahedral site is

$$r = \frac{a}{2} - \frac{a\sqrt{2}}{4} = 3.525 \times 10^{-8} \text{ cm} \left(\frac{1}{2} - \frac{\sqrt{2}}{4} \right) = 0.516 \times 10^{-8} \text{ cm}$$

19-26 Take the host atom radius to be r_1 and the interstitial atom radius to be r_2. Then from Table 19.2 $r_1 = \frac{\sqrt{3}}{4}a$, where a is the lattice parameter (the side length of the cubic unit cell).

An interstitial at the center of a face could come in contact either with the atom at the center of the cell, or with the four atoms at the corners of the face. The first possibility requires

$$2r_1 + 2r_2 \leq a$$

and the second requires that

$$2r_1 + 2r_2 \leq \sqrt{2}a$$

where we have used the fact that the face diagonal is $\sqrt{2}a$. Clearly the first condition is the stronger one, so the maximum interstitial radius is

$$r_2 = \frac{a}{2} - r_1 = \left(\frac{1}{2} - \frac{\sqrt{3}}{4} \right) a$$

The radius ratio is

$$\frac{r_2}{r_1} = \frac{4}{\sqrt{3}} \left(\frac{1}{2} - \frac{\sqrt{3}}{4} \right) = \frac{2}{\sqrt{3}} - 1 = 0.155$$

19-28 (a) Rb is metallic.   (b) C_5H_{12} is molecular.   (c) B is covalent.   (d) Na_2HPO_4 is ionic.

19-30 The nearest neighbors of a Na^+ ion are 6 Cl^- ions. The second nearest neighbors are a set of 12 Na^+ ions. The third nearest neighbors are a set of 8 Cl^- ions.

19-32

$$\Delta E_{diss} = \frac{1.7627 \times (1.602 \times 10^{-19} \text{ C})^2 \times 6.022 \times 10^{23} \text{ mol}^{-1}}{4\pi \times 8.854 \times 10^{-12} \text{ C}^2 \text{ J}^{-1} \text{ m}^{-1} \times 3.48 \times 10^{-10} \text{ m}}$$

$$= 7.04 \times 10^5 \text{ J mol}^{-1} = 704 \text{ kJ mol}^{-1}$$

Assuming a 10% reduction to approximate the repulsive energy we have

$$\Delta E = 633 \text{ kJ mol}^{-1}$$

19-34 (a)

$KBr(s) \rightarrow K(s) + 1/2Br_2(l)$	$\Delta E \approx -\Delta H_f^{\circ}(KBr(s)) = 393.80 \text{ kJ}$
$1/2Br_2(l) \rightarrow Br(g)$	$\Delta E \approx \Delta H_f^{\circ}(Br(g)) = 111.88 \text{ kJ}$
$K(s) \rightarrow K(g)$	$\Delta E \approx \Delta H_f^{\circ}(K(g)) = 89.24 \text{ kJ}$
$Br(g) + e^- \rightarrow Br^-(g)$	$\Delta E = -EA(Br) = -324.7 \text{ kJ}$
$K(g) \rightarrow K^+(g) + e^-$	$\Delta E = IE_1(K) = 418.8 \text{ kJ}$
$KBr(s) \rightarrow K^+(g) + Br^-(g)$	$\Delta E_{total} = 689.0 \text{ kJ}$

is obtained by adding the five equations. ΔE_{total} is the lattice energy of KBr. (Note: if we take account of the difference between ΔE and ΔH we find 684.1 kJ).

(b)

$$\text{lattice energy} = \frac{(1.7476)(1.602 \times 10^{-19} \text{ C})^2 (6.022 \times 10^{23} \text{ mol}^{-1})}{4\pi(8.854 \times 10^{-12} \text{ C}^2 \text{ J}^{-1} \text{ m}^{-1})(3.298 \times 10^{-10} \text{ m})}$$

$$= 7.361 \times 10^5 \text{ J mol}^{-1} = 736.1 \text{ kJ mol}^{-1}$$

which is slightly (7-8%) too large, because of the neglect of repulsive forces.

19-36 The presence of Schottky defects (vacancies) means that the measured density of a real crystal will inevitably be less than density of a hypothetical ideal crystal having no vacancies. This will cause estimates of Avogadro's number by the method of problem 19-15 and 19-16 to be high because the computation requires division by the density.

19-38 (a) The empirical formula of the nickel oxide is $Ni_{0.9796}O$. (The answer $NiO_{1.0208}$ is equivalent.) The formula is determined by comparing the relative number of moles of nickel and oxygen in a sample of any arbitrary size. Note that the mass percentage of oxygen is 21.77%.

(b) The set-up of the problem is just like the set-up in Example 19.6. The key equation is $3y + 2(0.9796 - y) = 2$ where y is the number of moles of nickel(III) per 1.000 mol of O. The value of y is 0.0408 mol, so the fraction of nickel in the +3 oxidation states is $0.0408/0.9796 = 0.0416$, or 4.16%.

19-40 Smectic phases involve the lining up of molecules in layers. One reason they might tend to form is if the center of one molecule is strongly attracted to the center of another. By lying in a single plane, the centers have a stronger attraction for one another, and lower energy.

19-42 The Bragg law is written for each of the two experiments, the first with x-rays of wavelength $\lambda_1 = 1.54$ Å and the second with x-rays of unknown wavelength λ_2. The ratio of the second to the first is:

$$\frac{n_1 \lambda_1}{n_2 \lambda_2} = \frac{d_1 \sin \theta_1}{d_2 \sin \theta_2}$$

But $n_1 = n_2$, $d_1 = d_2$, and λ_1 and θ_1 are given. It follows that $\lambda_2 = 1.65$ Å.

19-44 In x-ray diffraction, the greatest possible scattering angle 2θ is 180°, which corresponds to direct back-scattering. When 2θ is 180°, θ is 90° and $\sin \theta$ is 1.00. We substitute this into the Bragg law with $n = 1$ and $d = 4.20$ Å. The value of λ is 8.40 Å. If the wavelength of the probing x-rays exceeds the dimension of the feature being probed (the distance between the parallel planes) by more than a factor of two, no diffraction can occur.

19-46 (a) Substitute the given data into the Bragg law and compute $d = 3.344$ Å.

(b) The volume of the unit cell of polonium is d^3 because the distance between the parallel faces of the unit cell is the cell edge. The contents of the unit cell are a single Po atom because there is only one lattice point per cell and there must be an atom of Po for every lattice point. Also, the mass of a Po atom is 209 u. The density of Po is therefore $209 \text{ u}/(3.344 \text{ Å})^3 = 5.59$ u Å^{-3}. This converts to 9.28 g cm^{-3}. The density tabulated in Appendix F is close: 9.32 g cm^{-3}.

19-48 The problem tells us the number of atoms of each kind in the unit cell. The unit cell is the repeating motif of the crystalline substance, so its contents have the same ratio of atoms as the substance. There is 1 O atom per unit cell (1/8 atom per corner times 8 corners), 1 Cl atom (entirely within the cell), and there are 3 Na atoms (1/2 atom per face times 6 faces). The formula is Na_3ClO. This substance indeed exists; the answer is not a mix-up with sodium chlorate.

19-50

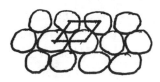

Disk radius $= r$

Unit cell (shown) is a rhombus with side length $2r$ and one disk per unit cell.

Area of sphere $= \pi r^2$

Area of rhombus $= 2r(2r \sin 60°) = 4\frac{\sqrt{3}}{2}r^2 = 2\sqrt{3}r^2$

Packing fraction $= \dfrac{\pi r^2}{2\sqrt{3}r^2} = 0.907$

This is just a cross section through the cylindrical fibers which must therefore have the same packing fraction.

19-52 The shortest Cl—Cl distance in LiCl is

$$\sqrt{2}(2.570 \text{ Å}) = 3.63 \text{ Å}$$

This is very close to twice the Cl^- ion radius from Appendix F (1.81 Å). In other words, for the LiCl distance to become any shorter in the rock salt structure, the Cl^-—Cl^- distance would become small enough that it would be less then the sum of the ionic radii. This would be prohibitive energetically. No such constraint of Cl^-—Cl^- repulsions applies to isolated LiCl molecules.

In addition, the bond in LiCl(g) is partially covalent, increasing the bond strength and reducing its length.

19-54 The difficult aspect of this problem is that it involves comparing both two different lattice types for a given element (e.g., CuI and CuI_2) and two different compounds for a given lattice type (e.g., CuI and CuBr). Let us write the overall reaction as

$$CuX_2(s) \rightarrow CuX(s) + \frac{1}{2}X_2(s,\ l,\ \text{or}\ g)$$

where X stands, as usual, for a halogen element and where the X_2 produced is gas, liquid or solid depending on the choice of X.

Stability is related to free energy, which in turn involves both enthalpy and entropy effects. Because iodine is solid under standard conditions while bromine is liquid and chlorine and fluorine are gases, the entropy charge should favor the above reactons for the *lighter* halogens. However, the problem states that the reaction is spontaneous only for the *heaviest* halogen, X = I. Thus enthalpy, not entropy, must determine the relative stabilities. Because ΔH is very close to the energy change ΔE, we will analyze the latter in a Born-Haber type of cycle taking reactants to products. It consists of the following steps:

$$CuX_2(s) \rightarrow Cu^{2+}(g) + 2X^-(g) \qquad \Delta E_1$$
$$Cu^{2+}(g) + e^- \rightarrow Cu^+(g) \qquad \Delta E_2$$
$$X^-(g) \rightarrow X(g) + e^- \qquad \Delta E_3$$
$$X(g) \rightarrow \tfrac{1}{2}X_2(g) \qquad \Delta E_4$$
$$\tfrac{1}{2}X_2(g) \rightarrow \tfrac{1}{2}X_2(g,\ l,\ \text{or}\ s) \qquad \Delta E_5$$
$$Cu^+(g) + X^-(g) \rightarrow CuX(s) \qquad \Delta E_6$$

It is easily verified that these add to give the correct overall equation. Let us examine the sign of the different ΔE's, and the effect on their magnitudes of a change in X.

Clearly ΔE_2 is large and negative, but X does not enter in and this is therefore constant. ΔE_3 is positive (equal to the election affinity of X). It is larger (more positive) for F, Cl, and Br, than for I; this is a trend in agreement with the observed stability difference. ΔE_4, however, shows a trend in the opposite direction. It is less negative for X = I than for the other halogens (because of the weakness of the I$_2$ bond). ΔE_5 is negative for X = I because of the attractions between I$_2$ in the solid; it is less negative for X = Br and zero for X = Cl or F. This trend is in agreement with observed stability.

This leaves the two lattice energy contributions: ΔE_1 is positive and ΔE_6 negative and almost certainly smaller in magnitude because of the smaller charge on copper. Moreover, the larger the anion the larger the interionic distance and the smaller the magnitude of the lattice energy. This trend therefore leads to lower stability for the CuI$_2$ compared to the other halides.

Overall, then, three factors favor the experimentally observed trend; the lower electron affinity of I, the greater attractions between I$_2$ molecules in the solid state, and the larger size of I$^-$ which reduces the magnitude of its lattice energies. The magnitudes of the various energy changes can be estimated from data in Appendix F, Table 3.2, and Appendix D. It seems likely that the last effect (the change in lattice energy with ionic size) is the largest one.

19-56 (a)

$$D = D_0 \exp\left[-E_a/RT\right]$$

$$= (0.145 \text{ cm}^2 \text{ s}^{-1}) \exp \left[-\frac{42220 \text{ J mol}^{-1}}{(8.315 \text{ J mol}^{-1} \text{ K}^{-1})(370.95 \text{ K})} \right]$$

$$= 1.65 \times 10^{-7} \text{ cm}^2 \text{ s}^{-1}$$

(b)

$$\overline{\Delta x^2} = 6Dt = 6(1.65 \times 10^{-7} \text{ cm}^2 \text{ s}^{-1})(60 \times 60 \text{ s})$$

$$(\overline{\Delta x^2})^{\frac{1}{2}} = 6.0 \times 10^{-2} \text{ cm} = 0.60 \text{ mm}$$

19-58 (a) covalent   (b) covalent within the chains, but molecular between chains (c) partially covalent along long chains of alternating atoms (—Si—O—Si—). The chains are themselves negatively charged and bonded ionically to positive ions.   (d) metallic.

19-60 The octahedral site will give rise to an octahedral crystal field splitting of the d-orbitals on the Cr^{3+}. Because Cr^{3+} is a d^3 species, its electron configuration will be $(t_{2g})^3$, with a single electron in each of the lower-energy, t_{2g}, orbitals.

19-62 The violet color of the fluorite shown must arise from absorption of the complementary color, yellow. From Fig. 15.3, the maximum absorption should be near 590 nm.

Chapter 20

Chemical Processes for the Recovery of Pure Substances

20-2 By-products are not the main product of economic value of a process, but nevertheless have some value. Waste materials have no value or negative value (if they present a disposal problem).

20-4 Biochemical processes can be mapped without difficulty onto a process diagram of the type shown in Figure 20.1. The biochemical production of sulfur, for example, has methane, carbon dioxide, and calcium sulfate as starting materials that undergo a chemical transformation by the intermediacy of marine bacteria to give the "desired product" sulfur and the by-products calcium carbonate and water. The methane itself (in natural gas) arises from the decay of the remains of marine organisms (along with petroleum). This too should be shown in the process diagram. Biochemical processes differ from processes in the chemical industry (with rare exceptions) in taking place at room pressure and near to room temperature, in employing enzymes instead of catalysts, and in requiring an aqueous medium. Industrial processes tend to use gas-phase reactions or heterogeneous reactions, and to employ extremes of T and P.

20-6 (a) Acid is added to water during its electrolysis to hydrogen and oxygen to increase the conductivity of the water and in that way increase the rate of production of the desired gases.

(b) One proof would be to note the quantity of sulfuric acid and continue the electrolysis to the point where the mass of hydrogen and the mass of oxygen exceed the mass of the hydrogen and oxygen in the sulfuric acid. Another would be to analyze for the concentrations of sulfate and hydronium ions from time to time during the electrolysis and compute the amounts of these species in the solution as functions of time, showing that they remain constant. A third way would be to substitute NaCl for H_2SO_4 and show that the electrolysis works the same.

20-8 Assume that $\Delta H°$ and $\Delta S°$ of the shift reaction are independent of temperature. Because $\Delta G = \Delta H° - T\Delta S°$ at constant temperature and pressure, it follows when ΔG is set equal to zero that $T = \Delta H°/\Delta S°$. It is routine to compute the $\Delta S°$ and $\Delta H°$ of a reaction from the $\Delta H_f°$ and $\Delta S°$ data in Appendix D; $\Delta H°$ is -41.17 kJ and $\Delta S°$ is -42.08 J K^{-1} for the "shift reaction" in this problem. Substitution gives the required temperature, 978 K.

20-10 (a) The equation for the reduction of the iron oxide of problem 18-9 with carbon monoxide is:

$$Fe_3O_4(s) + 4\,CO(g) \rightarrow 3\,Fe(s) + 4\,CO_2(g)$$

(b) The $\Delta H°$ of this reaction is

$$\Delta H° = 4\,\Delta H_f°(CO_2(g)) - 4\,\Delta H_f°(CO(g)) - \Delta H_f°(Fe_3O_4(s))$$

$$= 4(-393.51) - 4(-110.52) - 1(-1118.4) = -13.6 \text{ kJ}$$

An increase in temperature would cause a decrease in reaction yield for this exothermic reaction. An increase in pressure would have little effect on equilibrium yield.

20-12 First compute the chemical amount of hydrogen that is used to saturate the oleic acid.

$$n = \frac{PV}{RT} = \frac{(1.00 \text{ atm})(4.20 \text{ L})}{(0.08206 \text{ L atm mol}^{-1} \text{ K}^{-1})(298 \text{ K})} = 0.172 \text{ mol}$$

The mass of the stearic acid is the mass of the oleic acid, 48.5 g, plus mass of the hydrogen that was added during the hydrogenation. The mass of the H_2 is 0.172 mol $\times$ 2.016 g mol^{-1} = 0.347 g. There is therefore 48.8 g of stearic acid. The molar mass of stearic acid ($C_{18}H_{36}O_2$) is 284.48 g mol^{-1}, so the chemical amount of the stearic acid is 0.172 mol. It is clear that one mole of H_2 was gained by the oleic acid for every one mole of stearic acid produced. The formula of the oleic acid is therefore the formula of stearic acid minus 2 H's: $C_{18}H_{34}O_2$. There is one double bond in oleic acid.

20-14 A balanced equation for the reduction of U_3O_8 with calcium is

$$U_3O_8(s) + 8\,Ca(s) \rightarrow 3\,U(s) + 8\,CaO(s)$$

The standard enthalpy change of this reaction is

$$\Delta H° = 8(-635) + 3(0) - 8(0) - 1(-3575) = -1505 \text{ kJ}$$

and the standard entropy change is

$$\Delta S° = 8(39.8) + 3(50.2) - 8(41.4) - 1(282.6) = -144.8 \text{ J K}^{-1}$$

Use the relationship $\Delta G° = \Delta H° - T\Delta S°$ with $T = 298.15$ K to compute $\Delta G° = -1462$ kJ for the above reaction. The reaction produces 3 mol of uranium, so the standard molar enthalpy, and standard molar free-energy changes are one-third as large: $\Delta H° = -502$ kJ mol^{-1}, and $\Delta G° = -487$ kJ mol^{-1}.

20-16 The decomposition reaction is

$$\text{CuO}(s) \rightarrow \text{Cu}(s) + 1/2\,\text{O}_2(g)$$

Calculate the $\Delta H°$ and $\Delta S°$ of this reaction from the data in Appendix D. The tabulated values are on a molar basis, so the entry for $\text{O}_2(g)$ must be multiplied by 1/2. The results are $\Delta H° = 157.3$ kJ and $\Delta S° = 93.03$ J K^{-1}. The ratio $\Delta H°/\Delta S°$ then gives the approximate temperature. It is 1690 K.

20-18 (a) $\Delta G° = 2(0) + 3(-394.36) - 2(-764.08) - 3(0) = +345.08$ kJ
The reaction is not spontaneous under standard conditions.
(b) $\Delta H° = 2(0) + 3(-393.51) - 2(-842.87) - 3(0) = +505.21$ kJ
$\Delta S° = 2(32.64) + 3(213.63) - 2(75.90) - 3(5.74) = +537.15$ J K^{-1}

At high enough temperatures the reaction should become spontaneous. The crossover temperature at which ΔG becomes zero is

$$T = \frac{\Delta H°}{\Delta S°} = \frac{505,210 \text{ J}}{537.15 \text{ J K}^{-1}} = 940 \text{ K}$$

20-20 According to its formula, chalcopyrite is 34.94% S by mass and 34.63% Cu. We will use these facts without writing any chemical equations.

(a) The mass of copper producible from 1.0 metric tons of chalcopyrite is $0.3463 \times (1.0 \times 10^3 \text{ kg}) = 3.5 \times 10^2$ kg.

(b) Similarly, the mass of sulfur in the metric ton of ore is 349.4 kg. This would form 698.1 kg of SO_2, which is 1.090×10^4 mol of SO_2. This chemical amount of SO_2 occupies 2.4×10^5 L at 0°C and 1 atm.

20-22 The values of $\Delta G°$ are 150.7 kJ for the reduction of a mole of iron(III) oxide by $\text{C}(s)$ with $\text{CO}_2(g)$ as a product, 330.7 kJ for the reduction by $\text{C}(s)$ with $\text{CO}(g)$ as a product, and -29.43 kJ for the reduction by $\text{CO}(g)$ with $\text{CO}_2(g)$ as a product. The best reducing agent under standard conditions is $\text{CO}(g)$.

20-24 At the cathode the potassium ion is reduced: $e^- + K^+ \rightarrow K$; at the anode, hydroxide ion is oxidized to oxygen: $2\,OH^- \rightarrow 1/2\,O_2 + H_2O + 2\,e^-$.

20-26 Seven days is 6.048×10^5 s. A steady current of 75,000 A for this period means that 4.536×10^{10} C passes through the circuit. Dividing by the Faraday constant gives the chemical amount of electricity passing through the circuit. It is 4.70×10^5 mol. Every 2 mol of electrons theoretically accounts for the deposition of 1 mol of Mg. The theoretical yield of Mg is therefore 2.35×10^5 mol, which is 5.7×10^6 g.

20-28 If the temperature is high enough to melt the vanadium, as seems likely, the equation is:
$$V_2O_5(s) + 5\,Ca(l) \rightarrow 2\,V(l) + 5\,CaO(s)$$

Thus, 5 mol of Ca is required for every 2 mol of V. The 20.0 kg of V is 0.393 kmol of V so 0.982 kmol of Ca is theoretically required. This is 39.3 kg of Ca.

20-30 A current of 1.5 A for 22 minutes means that 1.98×10^3 C, which is 0.0205 mol of electrons, passes the cell. Each mole of electrons deposits one mole of silver on the spoon; since the molar mass of silver is 107.87 g mol^{-1}, this is 2.214 g of silver. The volume of this silver is 0.211 cm^3. This volume of silver covers the 16 cm^2 of spoon surface to an average thickness of 0.013 cm.

20-32 The reaction when CaH_2 is added to water is

$$CaH_2(s) + 2\,H_2O(l) \rightarrow Ca(OH)_2(aq) + 2\,H_2(g)$$

According to this equation, 2 mol of H_2 is generated for every 1 mol of CaH_2 reacted. The 12.21 g of CaH_2 is 12.21 g/42.094 g mol^{-1} = 0.2901 mol; it generates 0.5801 mol of H_2. The volume is

$$V = \frac{nRT}{P} = 14.2 \text{ L}$$

20-34 As long as liquid CO_2 and gaseous CO_2 are in equilibrium inside the tank, and the temperature is constant, the pressure of $CO_2(g)$ in the tank remains constant. Removal of $CO_2(g)$ simply causes some of the $CO_2(l)$ to vaporize to replace it. As Figure 5.21 shows, the equilibrium vapor pressure of $CO_2(g)$ near room temperature is close to 64 atm. Therefore the pressure gauge is almost certainly in good condition; if it is broken, it is coincidentally giving a correct reading. A suitable way to measure the amount of $CO_2(g)$ in the vessel is to determine the mass remaining by subtracting the mass of the empty cylinder from the mass of the cylinder that still has gas in it.

20-36 The oxidation number of carbon in ethylene C_2H_4 is -2; in ethane C_2H_6, it is -3. Because C decreases its oxidation number, it is reduced. It follows that the hydrogen is oxidized; the hydrogen increases its oxidation number from 0 to $+1$. The reaction is

$$C_2H_4 + H_2 \rightarrow C_2H_6$$

20-38 (a) $2C(s) + O_2(g) \rightarrow 2CO(g)$
$\Delta H° = 2(\text{-}110.52)$ kJ mol^{-1} = -221.04 kJ
$\Delta S° = 2(197.56) - 2(5.74) - 205.03 = 178.61$ J K^{-1}
$\Delta G = -221.04 - 178.6 \times 10^{-3}$ T

(b) $2Fe(s) + O_2(g) \rightarrow 2FeO(s)$
$\Delta H° = 2 \times (-266.27) = -532.54$ kJ
$\Delta S° = 2 \times (57.49) - 2 \times 27.28 - 205.03 = -144.61$ J K^{-1}
$\Delta G = -532.54 + 144.61 \times 10^{-3}$ T

(c) $2Ni(s) + O_2(g) \rightarrow 2NiO(s)$
$\Delta H° = 2 \times (-239.7) = -479.4$ kJ
$\Delta S° = 2 \times (37.99) - 2 \times 29.87 - 205.03 = -188.79$ J K^{-1}
$\Delta G = -479.4 + 188.79 \times 10^{-3}$ T

(d) $\frac{4}{3}Al(s) + O_2(g) \rightarrow \frac{2}{3}Al_2O_3(s)$
$\Delta H° = \frac{2}{3} \times (-1675.7) = $ -1117.1 kJ
$\Delta S° = \frac{2}{3} \times (50.92) - \frac{4}{3} \times 28.33 - 205.03 = -208.9$ J K^{-1}
$\Delta G = -1117.1 + 208.9 \times 10^{-3}$ T

(b) The reduction of $NiO(s)$ to $Ni(s)$ by $C(s)$ is thermodynamically feasible at all temperatures above about 700 K

The reduction of $FeO(s)$ to $Fe(s)$ by $C(s)$ is thermodynamically fesible at all temperatures above about 970 K.

(c) Aluminum oxide (Al_2O_3) cannot be reduced to Al by $CO(g)$ at any temperature in the range given (298 K to 1798 K). Extrapolation of the ΔG plots indicates that the reduction should be thermodynamically feasible above about 2300 K.

20-40 $\Delta G° = 4(-137.15) - (-1344) = +795$ kJ

20-42 The volume of a sphere is:

$$V = \frac{4}{3}\pi r^3$$

where r is its radius. The volume of the top 1 km of the earth's crust is the volume of a sphere of radius 6370 km minus the volume of a sphere of radius (6370 - 1) km. It is:

$$V_{(top\ of\ crust)} = \frac{4}{3}\pi[6370^3 - 6369^3]\ km = 5.1 \times 10^8\ km^8$$

A km is 10^5 cm so a km^3 is 10^{15} cm^3. The volume of the top kilometer of crust is 5.1×10^{23} cm^3. Each cm^3 weighs 2.8 g. Accordingly, the mass of the top 1 km of crust is 2.8 g cm^{-3} times the volume in cm^3 or 1.43×10^{24} g. The text states that the crust is 8 percent Al. The mass of Al in the crust is therefore $0.08 \times 1.43 \times 10^{24}$ g which is 1.1×10^{23} g or 1.1×10^{20} kg.

20-44 The $\Delta G°$ of the reaction $Mg(s) + H_2O(l) \rightarrow H_2(g) + MgO(s)$ is -332 kJ. The negative sign shows that the reaction is spontaneous at room conditions. It has an even greater tendency to proceed under the non-standard conditions in which the partial pressure of the product hydrogen is less than 1 atm, which is the case if water is simply poured over magnesium. The reaction is however quite slow because ordinary bulk samples of $Mg(s)$ are passivated by a coating of $MgO(s)$.

20-46

$$55.5\ kg\ crude\ Cu \times \left(\frac{98.3\ kg\ Cu}{100\ kg\ Cu}\right) \times \left(\frac{1\ mol\ Cu}{0.063546\ kg\ Cu}\right) \times \left(\frac{2\ mol\ e^-}{1\ mol\ Cu}\right)$$

$$\times \left(\frac{96485\ C}{1\ mol\ e^-}\right) \times \left(\frac{1\ s}{2.00 \times 10^3\ C}\right) = 8.3 \times 10^4\ s$$

This is about 1 day.

Chapter 21

Chemical Processes Based on Sulfur, Phosphorus and Nitrogen

21-2 The main reaction in the Claus process is the reduction of sulfur dioxide by hydrogen sulfide to produce elemental sulfur and water

$$SO_2(g) + 2\,H_2S(g) \rightarrow 3\,S(s) + 2\,H_2O(g)$$

We determine $\Delta H°$ and $\Delta S°$ from the data in Appendix D:

$$\Delta H° = 3(0) + 2(-241.82) - 1(-296.83) - 2(-20.63) = -145.55 \text{ kJ}$$

$$\Delta S° = 3(31.80) + 2(188.72) - 1(248.11) - 2(205.68) = -186.63 \text{ J K}^{-1}$$

The reaction is exothermic and involves a decrease in entropy of the reacting system. To favor a high equilibrium yield of the products, we should run the reaction at the lowest temperature consistent with an acceptable rate. We should also force the formation of products with high pressure.

21-4 The combustion of sulfur gives mainly $SO_2(g)$ although some other oxides of sulfur form only in very small amounts. Collecting the gas from the combustion in hydrogen peroxide solution oxidizes the SO_2 to $H_2SO_4(aq)$:

$$SO_2(g) + H_2O_2(aq) \rightarrow H_2SO_4(aq)$$

The amount of H^+ in this dilute aqueous sulfuric acid can be quite accurately determined by an acid-base titration. Alternatively, the sulfate ion could be precipitated with $Ba^{2+}(aq)$ and weighed. This is probably a better way to determine the sulfur unless it is known for certain that no other acidic oxides can arise from the sample.

21-6 The overall equation is: $H_2S(g) + 2\,O_2(g) \rightarrow H_2SO_4(l)$.

21-8 $H_2S_2O_7(l) + H_2O(l) \rightarrow 2H_2SO_4(aq)$.

21-10 Phosphoric acid is converted to NaH_2PO_4 in one run and to Na_2HPO_4 in another by controlled neutralization with NaOH. "Controlled neutralization" means that amounts of NaOH are added sufficient to neutralize one mole hydrogen ions in one batch and two moles of hydrogen ions in another batch. The sodium hydrogen phosphate and sodium dihydrogen phosphate are then reacted:

$$NaH_2PO_4(s) + 2\,Na_2HPO_4(s) \rightarrow Na_5P_3O_{10}(aq) + 2\,H_2O(g)$$

The mixture has to be heated to at least 300°C, well above the boiling point of water.

21-12 HPO_3

21-14 We multiply the number of molecules of NO created per flash by the number of flashes per year. Dividing this by Avogadro's number gives the chemical amount of NO produced by lightning per year. This is also the chemical amount of N that is fixed per year. Multiplying this amount by the molar mass of N gives the mass of N that is fixed each year. The number of flashes per year is 3.15×10^9, so 3.15×10^{36} molecules of NO are produced. This is 5.24×10^{12} mol of NO, which contains 5.24×10^{12} mol of N. This amounts to 7×10^{13} g of nitrogen, or 7×10^7 metric tons.

21-16 If the calcium hydroxide and carbon dioxide are *not* recycled, then the overall equation for the cyanamide process is

$$CaCO_3(s) + 2\,C(s) + N_2(g) + 4\,H_2O(l) \rightarrow$$
$$2\,NH_3(g) + 2\,CO_2(g) + CO(g) + Ca(OH)_2(s)$$

The $\Delta H°$ corresponding to this equation is +374.4 kJ.

Recycling the $Ca(OH)_2$ means that the equation

$$Ca(OH)_2(s) + CO_2(g) \rightarrow CaCO_3(s) + H_2O(l)$$

is added to the preceding. The result is

$$2\,C(s) + N_2(g) + 3\,H_2O(l) \rightarrow 2\,NH_3(g) + CO_2(g) + CO(g)$$

The $\Delta H°$ of the second reaction is -113.15 kJ. Hence, $\Delta H°$ of the overall reaction is +261.24 kJ. The $\Delta H°$ per mole of $NH_3(g)$ produced is half of this: +130.62 kJ mol^{-1}.

21-18 The equilibrium constant for the production of ammonia from its elements at 600 K is readily estimated using ΔG_f° for $NH_3(g)$ and ΔH_f° with the van't Hoff equation. The answer is $K_{600} \approx 4 \times 10^{-3}$.

Such a catalyst would be useful indeed. It would allow the use of lower pressures to achieve the same equilibrium yields of ammonia. This would save considerably on the cost of the equipment for the process.

21-20 The answer requires the application of Le Chatelier's principle. The first step of the Ostwald process, the oxidation of ammonia by oxygen, is favored by low pressure because 10 moles of gas are formed from 9 moles of gas. The two subsequent steps are both driven to the right by high pressure.

21-22 The balanced equations are:
$$3\,Ni(s) + 8\,HNO_3(aq) \rightarrow 3\,Ni(NO_3)_2(aq) + 2\,NO(g) + 4\,H_2O(l)$$
$$Tl(s) + 4\,HNO_3(aq) \rightarrow Tl(NO_3)_3(aq) + NO(g) + 2\,H_2O(l).$$

21-24 (a) The Lewis structure of diimine is $H{:}\ddot{N}{=}\ddot{N}{:}H$.

(b) The conversion of hydrazine to diimine is an oxidation—the oxidation number of nitrogen increases from -2 to -1. An oxidizing agent is required.

21-26 (a) The oxidation number of nitrogen in NH_2OH is -1; this is computed taking oxygen as having a -2 oxidation number and hydrogen a $+1$ oxidation number.

(b) The disproportionation of hydroxylamine to N_2 and N_2H_4 at pH 14 has the standard potential difference $\Delta \mathcal{E}^\circ = -0.73 - (-1.59) = 0.86$ V. The positive potential difference means the reaction is spontaneous under standard conditions.

21-28 The balanced equation is $2\,N_2H_4(l) + N_2O_4(g) \rightarrow 4\,H_2O(l) + 3\,N_2(g)$
Then: $\Delta H^\circ = 4(-285.83) + 3(0) - 2(50.63) - 1(9.16) = -1253.74$ kJ

21-30 Because Cl is more electronegative than H, the lone pair electrons on the P atom will be held more closely in PCl_3 than in PH_3, so PCl_3 will be a weaker Lewis base (a poorer donor of electron pairs).

21-32 A practical sequence to make N_2O would be to prepare ammonia from hydrogen and nitrogen by the Haber process, to oxidize some of the ammonia to nitric acid by the Ostwald process and then to react the remaining ammonia (a base) with nitric acid to give ammonium nitrate. Heating ammonium nitrate gives N_2O in good yield.

21-34 (a) Sulfur tends to disproportionate spontaneously at pH 14 to give HS^- and $S_2O_3^{2-}$. The $\Delta \mathcal{E}^\circ$ of the reaction is $-0.51 - (-0.74) = 0.23$ V.

(b) According to the reduction potentials given in the problem $S(s)$ is oxidized to $S_2O_3^{2-}$ more readily than HS^- is oxidized to $S(s)$ at pH 14; the $S(s)$ is the stronger reducing agent under these conditions.

21-36 Lead was used for the walls of the chamber in the manufacture of sulfuric acid because $PbSO_4$ is an insoluble salt, and forms on the surface of the metal as a barrier to further reaction.

21-38 The problem is solved by a series of unit conversions:

$$\frac{5.51 \times 10^{10} \text{ kg H}_2\text{SO}_4 \times 0.0003}{98.08 \times 10^{-3} \text{ kg mol}^{-1}} \left(\frac{1 \text{ mol SO}_2}{1 \text{ mol H}_2\text{SO}_4} \right) (64.06 \times 10^{-3} \text{ kg mol}^{-1}) = 1.08 \times 10^7 \text{ kg SO}_2$$

21-40 The coke reduces the Na_2SO_4 according to the equation:

$$Na_2SO_4(s) + 2\,C(s) \rightarrow Na_2S(s) + 2\,CO_2(g)$$

It is also possible that carbon monoxide forms as a product:

$$Na_2SO_4(s) + 4\,C(s) \rightarrow Na_2S(s) + 4\,CO(g)$$

Combining the ΔH_f°'s of the reactants and products gives ΔH° of the first reaction as $+235.26$ kJ and ΔH° of the second as $+580.2$ kJ. Either way, the production of the sodium sulfide is endothermic.

21-42 In the wet-acid process, phosphate rock (mainly fluorapatite) is reacted with sulfuric acid to make phosphoric acid. The production of triple superphosphate fertilizer consists of reacting phosphoric acid with phosphate rock. The balanced equations for these processes are given in Section 21.2. We multiply the first equation by 7 and the second by 3 and add them together. The H_3PO_4 terms cancel out, and the resultant equation, written with smallest whole-number coefficients, is:

$$2\,Ca_5(PO_4)_3F(s) + 7\,H_2SO_4(aq) + 17\,H_2O(l) \rightarrow$$

$$7\,CaSO_4\cdot2H_2O(s) + 2\,HF(g) + 3\,Ca(H_2PO_4)_2\cdot H_2O(s)$$

This equation shows that 7/3 mol of gypsum is generated per mole of triple superphosphate.

21-44 (a) According to the text, 5 mol of gypsum is produced for every 3 mol of phosphoric acid in the wet-acid process. Daily production in a fertilizer plant of 100 metric tons of phosphoric acid gives by-product according to:

$$100 \text{ tons} \times \left(\frac{10^6 \text{ g}}{1 \text{ ton}} \right) \times \left(\frac{1 \text{ mol H}_3\text{PO}_4}{98.0 \text{ g H}_3\text{PO}_4} \right) \times \left(\frac{5 \text{ mol CaSO}_4\cdot2\text{H}_2\text{O}}{3 \text{ mol H}_3\text{PO}_4} \right)$$

$$\times \left(\frac{172.2 \text{ g CaSO}_4 \cdot 2\text{H}_2\text{O}}{1 \text{ mol CaSO}_4 \cdot 2\text{H}_2\text{O}} \right) = 2.93 \times 10^8 \text{ g CaSO}_4 \cdot 2\text{H}_2\text{O}$$

This is the *daily* production of gypsum. If the plant operates 365 days a year, its annual production is 1.07×10^{11} g which is 107,000 metric tons.

(b) The volume of the gypsum is its mass divided by its density. Dividing the mass from part (a) by 2.32 g cm^{-3} gives a volume of 4.61×10^{10} cm^3, which equals 4.61×10^4 m^3. This would be a cube of gypsum about 36 meters on an edge.

21-46 The platinum is catalyzing the oxidation of ammonia. Observation of brown fumes means that $NO_2(g)$ is formed. One possible reaction is

$$4\,NH_3(g) + 7\,O_2(g) \rightarrow 4\,NO_2(g) + 6\,H_2O(g)$$

More likely is a reaction between NH_3 and O_2 to give $NO(g)$, which then reacts with $O_2(g)$ in a separate step to generate the $NO_2(g)$. The platinum foil stays hot (red-hot, in fact) as long as the reaction continues, being continually heated by the burning of the ammonia at its surface.

21-48 We assume that the $NO(g)$ behaves ideally and use the ideal-gas equation to calculate the chemical amount of $NO(g)$ that is generated. We have $V = 0.01146$ L, $P = 0.965$ atm, and $T = 291.95$ K, so $n_{NO} = 4.616 \times 10^{-4}$ mol. According to the balanced equation, 2 mol of nitrate reacts to form 2 mol of nitrogen oxide. The concentration of the nitrate is the chemical amount of nitrate divided by the volume of the solution: 4.616×10^{-4} mol/0.357 L $= 1.29 \times 10^{-3}$ M.

21-50 The observation of a non-zero dipole moment for hydrazine definitely eliminates all centrosymmetrical structures such as:

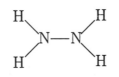

21-52 Strength:

$$OP(OH)_3 \approx O_2P_2(OH)_4 < O_3P_2(OH)_4 < O_5P_3(OH)_5 < O_6P_3(OH)_3.$$

The formulas have been re-written to highlight the ratio of the number of lone oxygen atoms to phosphorus atoms.

Chapter 22

Chemical Processes Based on the Halogens and the Noble Gases

22-2 The proper approach is to calculate $\Delta G°$ of the reaction from the standard free energies of formation and then to get K using $\Delta G° = -RT \ln K$. From the balanced equation and the data in Appendix D, we have

$$\Delta G° = 1(0) + 1(-228.1) + 6(-237.18) - 2(-131.23) - 4(-237.18) - 1(-465.17)$$

Adding up the numbers gives $\Delta G°$ equal to $+25.2$ kJ. The equilibrium constant is 3.9×10^{-5}.

22-4 The net ionic equation for the Solvay process is

$$2\,Na^+(aq) + CaCO_3(s) \rightarrow Na_2CO_3(s) + Ca^{2+}(aq)$$

The standard enthalpy change of this reaction is

$$\Delta H° = 1(-1130.68) + 1(-542.83) - 1(-1206.92) - 2(-240.12) = +13.6 \text{ kJ}$$

Overall, the Solvay process is slightly endothermic.

22-6 The half reactions are
Anode: $2Cl^-(aq) \rightarrow Cl_2(g) + 2e^-$
Cathode: $2H_2O(l) + 2e^- \rightarrow H_2(g) + 2OH^-(aq)$
$\Delta \mathcal{E}° = \mathcal{E}°(H_2O/H_2, OH^-) - \mathcal{E}°(Cl_2/Cl^-) = -0.8277 - 1.3583 = -2.186$ V
The minimum external voltage is 2.186 V.

22-8

$$I_2(s) \rightarrow I_2(aq)$$

$$\frac{0.33 \text{ g L}^{-1}}{253.8 \text{ g mol}^{-1}} = 1.3 \times 10^{-3} \text{ M} = [I_2] = K$$

$$\Delta G^\circ = -RT \ln K = -8.315 \text{ J mol}^{-1} \text{ K}^{-1} \times 298 \text{ K} \times \ln(1.3 \times 10^{-3})$$

$$= 16.5 \times 10^3 \text{J mol}^{-1} = 16.5 \text{ kJ mol}^{-1}$$

Appendix D gives $\Delta G^\circ = 16.40$ kJ mol^{-1}

22-10

$$3 \text{ BrO}^-(aq) \rightleftharpoons 2 \text{ Br}^-(aq) + \text{BrO}_3^-(aq)$$

$$\Delta H^\circ = 2 \times (-122) - 67 - 3 \times (-94) = -29 \text{ kJ}$$

$$\Delta S^\circ = 2 \times 82 + 162 - 3 \times 42 = +200 \text{ J K}^{-1}$$

$$\Delta G^\circ = \Delta H^\circ - T\Delta S^\circ = -29 \times 10^3 \text{ J} - (298 \text{ K})(+200 \text{ J K}^{-1}) = -88.6 \times 10^3 \text{ J}$$

$$K = \exp\left(-\frac{\Delta G^\circ}{RT}\right) = e^{-(-88.6 \times 10^3)/8.315 \times 298} = e^{+35.8} = 3.4 \times 10^{15}$$

22-12 (a) $3 \text{ Cl}_2(aq) + \text{I}^-(aq) + 6 \text{ OH}^-(aq) \rightarrow 6 \text{ Cl}^-(aq) + \text{IO}_3^-(aq) + 3 \text{ H}_2\text{O}(l)$
(b)

$$n_{\text{Cl}_2} = \frac{0.216 \text{ L} \times 1.00 \text{ atm}}{0.08206 \text{ L atm mol}^{-1} \text{ K}^{-1} \times 273.15 \text{ K}} = 9.64 \times 10^{-3} \text{ mol}$$

$$[\text{I}^-] = \frac{9.64 \times 10^{-3} \text{ mol}}{3 \times 0.0500 \text{ L}} = 0.0642 \text{ M}$$

22-14 The hypochlorite ion is reduced in the first half-reaction and oxidized in the second. The standard potential for it to disproportionate is $\Delta \mathcal{E}^\circ = 0.90 - 0.48 = 0.42$ V. This reaction is spontaneous.

22-16 (a) $\text{Br}_2(aq)$ can disproportionate into $\text{Br}^-(aq)$ and $\text{BrO}^-(aq)$ because the respective half-cell values are 1.06 V and 0.45 V. $\text{BrO}^-(aq)$, in turn, disproportionates into $\text{BrO}_3^-(aq)$ and additional $\text{Br}^-(aq)$. The stable products are thus $\text{BrO}_3^-(aq)$ and $\text{Br}^-(aq)$.

(b)
$$\text{BrO}_3^-(aq) + 3\text{H}_2\text{O}(l) + 6e^- \rightarrow \text{Br}^-(aq) + 6\text{OH}^-(aq)$$

$$\mathcal{E}^\circ = \frac{n_1 \mathcal{E}_1^\circ + n_2 \mathcal{E}_2^\circ}{n_1 + n_2} = \frac{4 \times (0.54\text{V}) + 2 \times (0.76 \text{ V})}{4 + 2} = 0.61 \text{ V}$$

22-18 Information on the formulas of products appears in the body of the chapter.
(a) $2\text{NaCl}(s) + \text{F}_2(g) \rightarrow 2\text{NaF}(s) + \text{Cl}_2(g)$
(b) $\text{MgO}(s) + \text{F}_2(g) \rightarrow \text{MgF}_2(s) + 1/2\,\text{O}_2(g)$
(c) $2\,\text{Na}(s) + \text{F}_2(g) \rightarrow 2\,\text{NaF}(s)$

22-20 The graphite fluoride contains 2.53 g/12.01 g mol^{-1} = 0.2106 mol C for every 1.000 g/19.00 g mol^{-1} = 0.0526 mol of F. The ratio of these two chemical amounts is 4.001 to 1, so the chemical formula is C_4F.

22-22 The molar mass of the compound is 102.45 g mol^{-1}. Therefore, 225 g of it is 2.196 mol. One mole of an (ideal) gas occupies 22.4 L at 0°C and 1.00 atm, so 2.196 mol occupies 49.2 L.

22-24 (a) The silicon has a steric number of 6; the ion is octahedral about the central Si.
(b) The central chlorine has a steric number of 4; the molecular ion is bent.
(c) The central iodine has a steric number of 5. The structure is see-saw.
(d) The central As has a steric number of 6; the molecular ion is octahedral.
(e) The central B has a steric number of 4; the molecular ion is tetrahedral.
(f) The central Br has a steric number of 6; the molecular ion is octahedral.

22-26 (a) The steric number of the sulfur is 4. The VSEPR theory predicts a tetrahedral geometry about the central sulfur atom in sulfuryl difluoride.

(b) The thionyl difluoride donates an electron pair to an oxygen atom. It acts as a Lewis base.

22-28 Reasoning by analogy to the case of water, we predict that the entropy of vaporization of BrF_3 will be *higher* than what is predicted by Trouton's rule.

22-30 (a) NaF (b) AsF_3 (c) C_5F_{12}

22-32 The reaction is $CF_4(g) + 2\,H_2O(l) \rightarrow CO_2(g) + 4\,HF(aq)$ and we have

$$\Delta G° = 4(-296.82) + (-385.98) - 2(-237.18) - 1(-879) = -220 \text{ kJ}$$

The reaction is spontaneous at standard conditions, but slow.

22-34 (a) The standard enthalpy change of the reaction in the problem is equal to the sum of the standard enthalpies of formation of the products minus those of the reactants. Let x equal $\Delta H_f°$ for $XeF_4(g)$. Then

$$\Delta H° = -887 \text{ kJ} = 1 \underbrace{(0)}_{Xe(g)} + 4 \underbrace{(-271.1)}_{HF(g)} - 1 \underbrace{(x)}_{XeF_4(g)} - 1 \underbrace{(0)}_{H_2(g)}$$

Solving gives $x = -197$ kJ mol^{-1}.

(b) We know from part (a) that $\Delta H_f°$ of $XeF_4(g)$ is -197 kJ mol^{-1}; hence for the reaction

$$Xe(g) + 2\,F_2(g) \rightarrow XeF_4(g)$$

we have $\Delta H° = -197$ kJ. This formation reaction can be imagined as proceeding with the production of 4 mol of $F(g)$ from $F_2(g)$ and then the creation of 4 mol of Xe—F bonds. It costs 4(79 kJ) to create $4\,F(g)$ from $2\,F_2(g)$. This means that the formation of the 4 mol of Xe—F bonds is exothermic to the extent of 513 kJ. The mean bond enthalpy is one-fourth of this, which is 128 kJ.

22-36 (a) The $\Delta E_f°$ is the heat absorbed by a reaction at constant volume that goes to give a compound in its standard state from its constituent elements in their standard states. The $\Delta E_f°$ of 2.763×10^{-4} mol of $XeO_3(s)$ is therefore $+112$ J. The formation of one mole of the compound would absorb 405 kJ, that is, $\Delta E_f° = 405$ kJ mol^{-1}.

(b) The best way to answer the question is to write the relationship $H = E + PV$. From this it follows that $\Delta H = \Delta E + \Delta(PV)$ and $\Delta H = \Delta E + (\Delta n_g)RT$ where Δn_g is the change in the chemical amount of gas and it is assumed that the volumes of the solids and liquids are negligible and the temperature is constant. In the formation reaction, Δn_g is negative, so $\Delta H_f° < \Delta E_f°$.

22-38 (a) Perxenic acid is H_4XeO_6. Its large positive standard reduction potential means it is an excellent oxidizing agent.

(b) The XeO_3 is itself an excellent oxidizing agent, as shown by the large positive standard reduction potential of the second half-reaction. It is a very poor reducing agent.

(c) We calculate $\Delta\mathcal{E}°$ from the $\mathcal{E}°$'s of the half-reactions in which the $XeO_3(aq)$ is reduced and oxidized. The result is $\Delta\mathcal{E}° = 2.12 - 2.36 = -0.24$ V. Hence 1 M $XeO_3(aq)$ is stable with respect to disproportionation.

22-40 (a)

$$2\,HCl(g) + \frac{1}{2}\,O_2(g) \rightleftharpoons Cl_2(g) + H_2O(g)$$

$$\Delta H° = -241.82 - 2(-92.31) = -57.20 \text{ kJ}$$

$$\Delta S° = 188.72 + 222.96 - 2 \times (186.80) - \frac{1}{2} \times 205.03 = -64.44 \text{ J K}^{-1}$$

$$\Delta G = -57.20 \times 10^3 - 723.15 \times (-64.44) = -10.60 \times 10^3 \text{ J}$$

$$K = e^{10.60 \times 10^3 / 8.315 \times 723.15} = e^{1.76} = 5.8$$

(b) The HCl bond is stronger than that in HBr by 66 kJ mol^{-1} and two bonds must be broken for each Cl_2 formed, costing 132 kJ mol^{-1}. This is only partially compensated for by the fact that the bond in Cl_2 is 50 kJ mol^{-1} stronger than that in Br_2. The net effect is that HBr will react more readily than HCl with O_2, and the equilibrium constant will be larger.

22-42

$$\frac{0.033 \text{ g}}{(0.100 \text{ L})(253.8 \text{ g mol}^{-1})} = 1.3 \times 10^{-3} \text{ M}$$

For the equilibrium $I_2(s) \rightleftharpoons I_2(aq)$ we have then $K = [I_2] = 1.3 \times 10^{-3}$. For the second equilibrium, $I^-(aq) + I_2(aq) \rightleftharpoons I_3^-(aq)$ we have, using Appendix D,

$$\Delta G^\circ = -51.4 - (-51.57) - 16.40 = -16.2 \text{ kJ}$$

$$\ln K = -\frac{\Delta G^\circ}{RT} = \frac{16,200 \text{ J mol}^{-1}}{(8.315 \text{ J mol}^{-1}\text{K}^{-1})(298 \text{ K})} = 6.55$$

$K = 700$

$$\frac{[I_3^-]}{[I^-][I_2]} = 700 = \frac{[I_3^-]}{[I^-](1.3 \times 10^{-3})}$$

because in contact with $I_2(s)$ we must have $[I_2] = 1.3 \times 10^{-3}$ M.

$$\frac{[I_3^-]}{[I^-]} = 0.91; \quad [I^-] = 1.0 - x; \quad [I_3^-] = x$$

$$\frac{x}{1.0 - x} = 0.91; \quad x = 0.48$$

Total solubility of $I_2 = [I_2(aq)] + [I_3^-(aq)] = 1.3 \times 10^{-3} + 0.48 = 0.48$ M. This is 120 g L^{-1}, or 12 g of I_2 per 100 g, assuming the density of the 1.0 M KI solution to be close to that of pure water.

22-44 Iodine lies near a diagonal band of semiconducting elements from silicon to tellurium. Thus it is not surprising that, like a semiconductor, it has a small conductivity that increases with temperature. Bromine should have a smaller conductivity because it lies farther from this diagonal band.

22-46 We compute the quantity of electricity needed to oxidize the fluoride ion:

$$3.3 \times 10^3 \text{ g F}_2 \times \left(\frac{1 \text{ mol F}_2}{38.0 \text{ g F}_2}\right) \times \left(\frac{2 \text{ mol } e^-}{1 \text{ mol F}_2}\right) \times \left(\frac{96,485 \text{ C}}{1 \text{ mol } e^-}\right) = 1.676 \times 10^7 \text{ C}$$

This much electricity passes in 1 hour, so the quantity per second is 1/3600 as much or 4.655×10^3 C s^{-1}. This means the current is 4.7×10^3 A.

22-48 (a) Assuming HF(g) to obey the ideal gas law, we expect

$$\frac{n}{V} = \frac{P}{RT} = \frac{1.00 \text{ atm}}{0.08206 \text{ L atm mol}^{-1} \text{ K}^{-1} \times 293.15 \text{ K}} = 0.0416 \text{ mol L}^{-1}$$

Predicted vapor density $= 0.0416$ mol $L^{-1} \times 20.0$ g mol$^{-1} = 0.831$ g L^{-1}

(b) The large negative deviation of the calculated vapor density from the observed vapor density is a consequence of its extensive association (hydrogen bonding) in the gaseous state.

22-50 The most obvious effect of $F^-(aq)$ in the body would be to tie up calcium in an insoluble compound: $Ca^{2+}(aq) + 2\,F^-(aq) \rightarrow CaF_2(s)$.

22-52 The reactions are

$$PuO_2 + 3\,O_2F_2 \rightarrow PuF_6 + 4\,O_2$$

$$Pu + 3\,O_2F_2 \rightarrow PuF_6 + 3\,O_2$$

22-54 (a) The molecule of S_2F_{10} consists of two sulfur atoms joined by a single bond with each S having 5 F's also single-bonded to it. We do not expect fluorine to form more than one bond.

(b) Both S's in this model of S_2F_{10} would have steric numbers of 6. The shape of the molecule would therefore be two octahedra bonded at a corner.

22-56 The ionization energy of $Xe(g)$ is 1170.4 kJ mol^{-1}, which is nearly equal to the ionization energy of $O_2(g)$.

22-58 In this reaction, the AsF_5 is a fluoride-ion acceptor, which means it is an electron-pair acceptor and hence an acid. The XeF_2 is an electron-pair donor and hence a base.

Chapter 23

Ceramic Materials

23-2 The Lewis dot structure of the cyclosilicate ion displays a total of 144 valence electrons. It consists of 6 Si and 6 O atoms alternating in a ring. Each Si atom has 2 other O's linked to it as side-groups on the ring. All bonds are single.

23-4 (a) Tremolite consists of infinite double chains. All O's are in the -2 oxidation state, and all Si's in the $+4$ state. The Mg and Ca are both $+2$ and the hydrogen is $+1$. The assignment to a structural type is made on the basis of the ratio of silicate oxygen to silicon, which is 2.75 to 1.

(b) Gillespite consists of infinite sheets. The Si and O are in the $+4$ and -2 oxidation states respectively (as they are in all the silicate minerals in this problem) and the Ba and Fe are both $+2$.

(c) Uvarovite contains silicate tetrahedra. The Ca is in the $+2$ state, and the Cr is in the $+3$ state.

(d) Barysilate contains pairs of silicate tetrahedra. The Mn and Pb are both in the $+2$ oxidation state.

23-6 (a) The aluminosilicate mineral amesite derives formally from a silicate with a 5 to 2 oxygen-to-silicon ratio. Based on this ratio it contains infinite sheets of aluminosilicate units. The aluminum is in the $+3$ oxidation state, the Si and O are in the $+4$ and -2 states, respectively. This is true in all the aluminosilicates in this problem. The Mg is in the $+2$ state and the "other aluminum" (that which is not part of the silicate framework) is $+3$ aluminum. The hydroxide hydrogen and hydroxide oxygen are $+1$ and -2 respectively.

(b) Phlogopite contains infinite sheets of aluminosilicate units. The K is $+1$ and the Mg is $+2$. Other atoms are as before.

(c) Thomsonite contains an infinite network of aluminosilicate units. The Na and Ca are $+1$ and $+2$ respectively, and the Al and Si are $+3$ and $+4$.

23-8 In acid solution ($HClO_4$):

$$Pb^{2+}(aq) + 2e^- \rightarrow Pb(s); \quad \mathcal{E}^\circ = -0.126 \text{ V}$$

In basic solution:

$$Pb^{2+}(aq) + 2\ OH^-(aq) \rightarrow PbO(s) + H_2O(l)$$

$$PbO(s) + H_2O + 2e^- \rightarrow Pb(s) + 2\ OH^-(aq) \quad \mathcal{E}^\circ = -0.576 \text{ V}$$

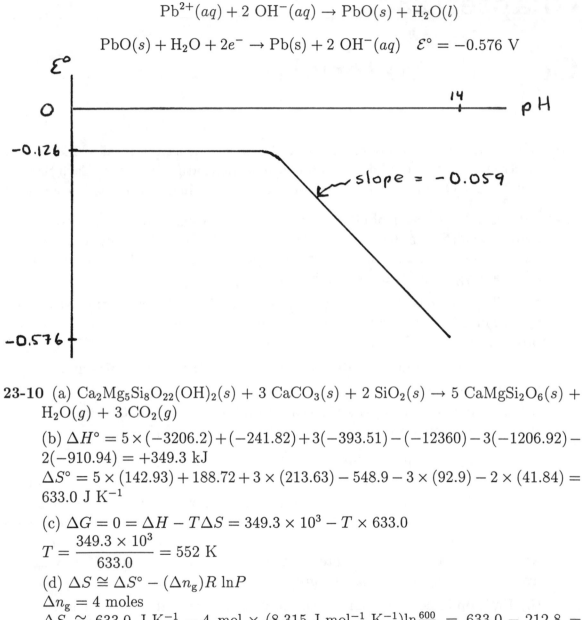

23-10 (a) $Ca_2Mg_5Si_8O_{22}(OH)_2(s) + 3\ CaCO_3(s) + 2\ SiO_2(s) \rightarrow 5\ CaMgSi_2O_6(s) + H_2O(g) + 3\ CO_2(g)$

(b) $\Delta H^\circ = 5 \times (-3206.2) + (-241.82) + 3(-393.51) - (-12360) - 3(-1206.92) - 2(-910.94) = +349.3$ kJ

$\Delta S^\circ = 5 \times (142.93) + 188.72 + 3 \times (213.63) - 548.9 - 3 \times (92.9) - 2 \times (41.84) = 633.0$ J K^{-1}

(c) $\Delta G = 0 = \Delta H - T\Delta S = 349.3 \times 10^3 - T \times 633.0$

$$T = \frac{349.3 \times 10^3}{633.0} = 552 \text{ K}$$

(d) $\Delta S \cong \Delta S^\circ - (\Delta n_g)R \ln P$

$\Delta n_g = 4$ moles

$\Delta S \cong 633.0$ J K^{-1} $- 4$ mol $\times$ (8.315 J mol^{-1} K^{-1})$\ln\frac{600}{1} = 633.0 - 212.8 = 420.2$ J K^{-1}

$$T \cong \frac{349.3 \times 10^3}{420.2} = 831 \text{ K}$$

23-12 (a) $\Delta H° = -909.48 - (-910.94) = +1.46$ kJ
$\Delta S° = 42.68 - 41.84 = +0.84$ J K^{-1}
$\Delta G° = -855.43 - (-856.67) = +1.24$ kJ

(b) Quartz has the lower free energy and is stable under standard conditions.

(c) Because cristobalite has the higher entropy, it is more stable at very high temperatures.

(d) High pressures favor the state of smaller molar volume, quartz.

(e) For the two phases to coexist at $P = 1$ atm, we must have

$$\Delta G = 0 \cong \Delta H° - T\Delta S°$$

$$\frac{\Delta H°}{T} = \Delta S°$$

$$\Delta V = 25.74 - 22.69 = 3.05 \text{ cm}^3 = 3.05 \times 10^{-6} \text{ m}^3$$

$$\left(\frac{dP}{dT}\right)_{\text{coex}} = \frac{\Delta H}{T\Delta V} \approx \frac{\Delta S°}{\Delta V} = \frac{0.84 \text{ J K}^{-1}}{3.05 \times 10^{-6} \text{ m}^3}$$

$$= 2.8 \times 10^5 \text{ J m}^{-3} \text{ K}^{-1} = 2.8 \times 10^5 \text{ Pa K}^{-1} = 2.7 \text{ atm K}^{-1}$$

23-14

$$\underbrace{3 \text{ Al}_2\text{Si}_4\text{O}_{10}(\text{OH})_2(s)}_{\text{pyrophyllite}} \rightarrow \underbrace{\text{Al}_6\text{Si}_2\text{O}_{13}(s)}_{\text{mullite}} + \underbrace{10 \text{ SiO}_2(s)}_{\text{cristobalite}} + 3 \text{ H}_2\text{O}(g)$$

23-16 The dehydration of kaolinite ($\text{Al}_2\text{Si}_2\text{O}_5(\text{OH})_4$) produces 2 mol of water per mol of the mineral. The molar mass of kaolinite is 258.16 g mol^{-1}. The chemical amount of kaolinite is its mass divided by its molar mass. Doubling this gives the chemical amount of steam, 31.0 mol. This quantity is inserted in the ideal-gas equation with $T = 873.15$ K and $P = 1$ atm. The answer is 2.2×10^3 L.

23-18 We imagine a 100-g sample of portland cement and compute the masses of the several oxides. From these we obtain the chemical amounts of the oxides by dividing each mass by the proper molar mass. The formulas of the oxides then allow computation of the chemical amounts of the six elements and of course of the total chemical amount of oxygen. We put this on a basis of 1.00 mol of O. The answer can be summarized in the "chemical formula" $\text{Ca}_{0.514}\text{Si}_{0.158}\text{Al}_{0.052}\text{Fe}_{0.016}\text{Mg}_{0.028}\text{S}_{0.010}\text{Na}_{0.020}\text{O}_{1.00}$.

23-20 The slaking of lime is represented:

$$CaO(s) + H_2O(l) \rightarrow Ca(OH)_2(s)$$

The $\Delta H°$ for this reaction is

$$\Delta H° = 1(-986.09) - 1(-635.09) - 1(-285.83) = -65.17 \text{ kJ}$$

The slaking of 1 mol (56.08 g) of $CaO(s)$ is exothermic to this extent. The slaking of 1.00 kg releases proportionately more heat, 1162 kJ. This is only an estimate because the product is unlikely to be solid $Ca(OH)_2$ when lime is slaked under ordinary conditions. Dissolution of some of the $Ca(OH)_2(s)$ releases more heat.

23-22 The sum of the oxidation states of the single atoms must equal zero. The oxidation states of Tl, Ca, Ba and O are +3, +2, +2 and −2 respectively. Let the oxidation state of Cu be y. Then:

$$(2 \times 3) + (2 \times 2) + (2 \times 2) + (3y) + (10.5 \times -2) = 0 \quad \text{and} \quad y = 7/3 = 2.33$$

23-24 (a) $BCl_3(g) + NH_3(g) \rightarrow BN(s) + 3\,HCl(g)$

(b) For this reaction:

$$\Delta H° = 1(-254.4) + 3(-92.31) - 1(-403.76) - 1(-46.11) = -81.46 \text{ kJ}$$

Thus, the $\Delta H°$ per mole of $BN(s)$ is $-81.46 \text{ kJ mol}^{-1}$.

(c) Boron nitride has extremely low electrical conductivity and a very high melting point. It is nearly as hard as diamond.

23-26 The oxidation of $B_4C(s)$ is represented: $B_4C(s) + 4\,O_2(g) \rightarrow 2\,B_2O_3(s) + CO_2(g)$. The $\Delta G°$ of this reaction is computed from $\Delta G_f°$'s in the usual way:

$$\Delta G° = 2(-1193.70) + 1(-394.36) - 4(0) - 1(-71) = -2711 \text{ kJ}$$

Boron carbide is thermodynamically unstable in oxygen at standard conditions and thermodynamically unstable in air as well. It reacts only slowly, however.

23-28 The $Si_{12}O_{30}^{12-}$ ion is six times the $Si_2O_5^{2-}$ unit. The structure should contain infinite sheets. The oxidation state of Si is +4, of O is −2, of H is +1, of Cl is −1, and of Mg is +2. Combining these gives a total oxidation number for the remaining $Mn_{12}Fe$ unit of +26. Thus both the Mn and the Fe must be in the +2 oxidation state.

23-30 The Na^+ ion and Ca^{2+} ion have nearly the same radius (0.98 and 0.99 Å). They substitute freely in each other's sites in solid solutions of albite and anorthite. The radius of the K^+ ion is greater (1.33 Å) and its substitution in sites in albite and anorthite is restricted.

23-32 The given quantities define the zeolite $K_2O \cdot Al_2O_3 \cdot 4(SiO_2) \cdot 6(H_2O)$. This compound is 9.91% Al by mass.

23-34 A small change dT in temperature and dP in pressure changes the free energy G in the liquid by

$$dG_l = V_l dP - S_l dT$$

where V_l is the volume and S_l the entropy of the liquid. The corresponding change for the solid is

$$dG_s = V_s dP - S_s dT$$

Along the coexistence curve, the free energies of liquid and solid remain equal, so

$$dG_l = dG_s$$
$$V_l dP - S_l dT = V_s dP - S_s dT$$
$$(V_l - V_s)dP = (S_l - S_s)dT$$
$$\left(\frac{dP}{dT}\right)_{coex} = \frac{S_l - S_s}{V_l - V_s} = \frac{\Delta S}{\Delta V} = \frac{\Delta H}{T \Delta V}$$

23-36 Soda-lime glass is an amorphous solid of approximate composition 73% SiO_2, 17% Na_2O, 5% CaO, and 5 % MgO. It contains little aluminum. Fired kaolinite contains a great deal of aluminum, having the approximate composition $Al_6Si_2O_{13}$. Soda-lime glass contains a random, three-dimensional ionic network $(SiO_3^{2-})_n$ the charge of which is locally neutralized by the Na^+, Ca^{2+} and Mg^{2+} ions. Fired kaolinite includes Al as part of structural aluminosilicate chains. Fired kaolinite has undergone an irreversible chemical reaction; its physical shape in unalterable without breakage. Glass can be melted and reformed.

23-38 Magnesia (MgO) is an O^{2-} donor and is a basic refractory. Silica is an O^{2-} acceptor and is an acidic refractory. The two react: $MgO + SiO_2 \rightarrow MgSiO_3$.

23-40 In acid: $BeO(s) + 2H_3O^+(aq) + H_2O(l) \rightarrow Be(H_2O)_4^{2+}(aq)$.

In base: $BeO(s) + 2OH^-(aq) + H_2O(l) \rightarrow Be(OH)_4^{2-}(aq)$.

23-42 The oxide ceramics are in general thermodynamically stable with respect to their constituent elements. The nonoxide ceramics are often *not* thermodynamically stable with respect to reaction to form oxides (as in air).

Chapter 24

Optical and Electronic Materials

24-2 In each case the maximum wavelength is calculated by setting the band-gap energy equal to hc/λ. For GaAs, it is 867 nm; for CdS, it is 512 nm. The former is in the infrared region of the spectrum; this latter is near the middle of the visible region of the spectrum.

24-4 The equation for the number of electrons excited to the conduction band per cubic centimeter in a semiconductor is

$$n_e = (4.8 \times 10^{15}\ \text{cm}^{-3}\text{K}^{-3/2})T^{3/2}e^{-E_g/2RT}$$

Substitution of the band gap energy with due regard for the cancellation of units gives $n_e = 4.3 \times 10^{13}\ \text{cm}^{-3}$ for germanium.

24-6 (a) Ge doped with In is a p-type semiconductor (b) CdS doped with As is a p-type semiconductor if As substitutes for S. If As substitutes at random for Cd or S then it raises the number of electrons in the product and makes it n-type.

24-8 We substitute the band gap energy of the LED into the equation $E = hc/\lambda$ to compute the required wavelength. It is 584 nm, in the yellow region of the visible spectrum.

24-10 The decrease in the band-gap energy that accompanies the conversion of cinnabar to metacinnabar is accompanied by a shift in the wavelength corresponding to excitation across the band gap of the pigment from 621 nm (in cinnabar) to 764 nm. The bright crimson of the cinnabar fades away. In the first case, red light with wavelengths longer than 621 nm is transmitted and the pigment appears red; in the second case, all colors of visible light are absorbed and the sample appears black.

24-12

$$E = \frac{hc}{\lambda} = 2.84 \times 10^{-19} \text{ J} = 171 \text{ kJ}$$

This is a little short of 5 times 34.5 kJ, so at most 4 molecules of ATP could be produced per photon.

24-14 The band gap becomes larger going from a metal to a semiconductor to an insulator.

24-16 The conductivity of the Sb-doped Si (an n-type semiconductor) should decrease as Ga is added to a minimum when the ratio of the dopants is 1 to 1. At this point the doubly-doped Si is electronically equivalent to pure silicon itself. More Ga beyond this point should make the conductivity increase, as the semiconductor becomes p-type

24-18 The energy should be lowered, because the two energy levels will split into an unoccupied upper level and an occupied lower level.

Chapter 25

Polymeric Materials

25-2 $n\,C_2F_4 \rightarrow (CF_2CF_2)_n$.

25-4 Polymethyl methacrylate forms by addition polymerization of methyl methacrylate, which has the structure:

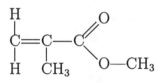

25-6 The starting monomer, which is alanine, has the structure:

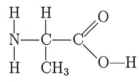

25-8 The repeating unit in the polyester has the formula $C_{10}H_8O_4$. This formula is the sum of the molecular formulas of terephthalic acid and ethylene glycol minus twice the formula of water, a relationship that derives from the fact that the diacid and diol combine by condensation with loss of water. The molar mass of the repeating unit is 192.17 g mol^{-1}. Then the number of moles of repeating unit needed is:

$$(10.0 \times 10^3) \text{ g polymer} \times \left(\frac{1 \text{ mol units}}{192 \text{ g polymer}} \right) = 52.04 \text{ mol}$$

This means that the synthesis needs 52.04 mol of terephthalic acid ($\mathcal{M} = 166.13$ g mol^{-1}) and 52.04 mol of ethylene glycol ($\mathcal{M} = 62.07$ g mol^{-1}). These

238

chemical amounts convert to 8.65 kg of terephthalic acid and 3.23 kg of ethylene glycol.

25-10 The mass of polystyrene is 2.84×10^{12} g, and the mass of a monomer unit of styrene (C_8H_8) is 104.2 g mol^{-1}. In addition polymerization, no mass is lost by the splitting out of small molecules. Hence, 2.73×10^{10} mol of styrene was incorporated, which is 1.64×10^{34} molecules (by multiplication by Avogadro's number).

25-12 The ring form of D-ribose has four asymmetric carbon atoms:

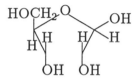

25-14 There are $20^{10} = 1.0 \times 10^{13}$ possible polyeptides containing 10 amino acids. There are $20^{100} = 1.3 \times 10^{130}$ possible polypeptides containing 100 amino acids.

25-16 The pentapeptide is:

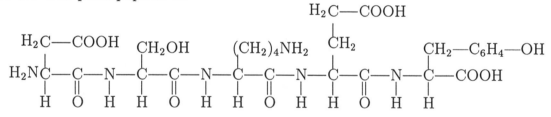

The pentapeptide has mostly polar side-groups. It should be more soluble in water than in *n*-octane.

25-18 A nucleotide consists of a molecule of deoxyribose ($C_5H_{10}O_4$), a molecule of phosphoric acid (H_3PO_4) and a molecule of one of the four bases (adenine ($C_5H_5N_5$), cytosine ($C_4H_5N_3O$), thymine ($C_5H_6N_2O_2$), and guanine ($C_5H_5N_5O$)) linked as a unit by the loss of two molecules of water. The average molar mass of a nucleotide is 327 g mol^{-1}. The links between nucleotides form with the loss of still another molecule of water, so the average molar mass of a repeating unit in DNA is 309 g mol^{-1}. If the molar mass of the DNA is 4×10^9 g mol^{-1}, there are 1.3×10^7 nucleotides along the DNA chain. Each link in DNA contains a base, so this DNA is said to have 1.3×10^4 kilobases. Of course this answer could change if the amounts of the four nucleotides are not approximately equal.

25-20 The polymerization reaction must be driven by an increase in entropy in the surroundings, because it itself involves a local decrease in entropy. Heat must therefore be evolved into the surroundings; the reaction is exothermic.

25-22 If the only difference between natural rubber and gutta-percha is the geometrical relationship (*cis* in natural rubber, but *trans* in gutta-percha) at the recurring double bonds, then conversion of the double bonds to single bonds by reaction with H_2 should give identical products.

25-24 The structure of L-sucrose is the mirror image of that in Figure 25.10b.

25-26 Suppose there is 1.00 mol of hemoglobin. This weighs 65,000 g and contains 223.6 g of Fe. This much Fe is 4.0 mol of Fe. Each mole of hemoglobin contains 4.0 mol of Fe, so each molecule of hemoglobin contains 4 atoms of Fe.

25-28 TGAACTGGC

Appendix A

A-2 The trailing zeros in (b), (c), and (e) must not be omitted when the number is put into scientific notation. The two negative signs in answer (b) sometimes trouble students.
(a) 4.579×10^3 (b) -5.020×10^{-2}
(c) 2.134560×10^3 (d) 3.825 (e) 4.50×10^{-5} (f) 9.814

A-4 (a) 0.003333 (b) $-12,000,000$ (c) 0.0000279 (d) 30 (e) 0.06700.

A-6 The amount is 4.6×10^{-6} g.

A-8 (a) Statistical methods for deciding whether to omit a outlier are not developed in the Appendix. Instead the appeal is to use good judgment. Although there is plenty of scatter, none of the values is grossly out of line (as in the preceding problem) and all should be retained.

(b) The average volume is 555 cm^3.

(c) The standard deviation is $\sigma = 27$ cm^3, and the confidence limit is ±20.

A-10 The average of the determinations of mass in problem A-7 differs by 6.4 % from the "true" value (the one obtained with better balance). The average volume in problem A-8 differs by only 0.36 % from the "true" volume. The second measurement was more accurate despite being less precise.

A-12 (a) two (b) ambiguous (one, two, three or four significant figures) (c) seven (d) four (e) three.

A-14 (a) -0.0025 in (no change needed) (b) 7.0×10^3 g (c) 1.4×10^2 s
(d) 2.7×10^7 Pa (e) 2.0×10^{-19} J

A-16 Eight, 2,997,215.5; seven, 2,997,216; six, 2.99722×10^6; five, 2.9972×10^6; four, 2.997×10^6; three, 3.00×10^6; two, 3.0×10^6; one, 3×10^6.

A-18 (a) 250.89 (b) -77.7 (c) 6.552×10^{19} (d) -1.467×10^{-13}

A-20 (a) 4190 (b) 0.257 (c) 2.948×10^{-26} (d) 6.8×10^{-4}

A-22 The length of the table is 198.88 in. Five significant figures appear in the answer (not three) because the value "2.54" is a definition and, despite its appearance, has an infinite number of significant figures, not three.

Appendix B

B-2 (a) 6.6×10^{-5} K   (b) 1.59×10^7 kg m^{2}s^{-2} (c) 1.3×10^{-7} kg
(d) 6.2×10^{10} kg m^{-1}s^{-2}

B-4 (a) The answer is 9032°F assuming four significant figures in the given temperature. It is very hard to measure such a high temperature to within one degree Celsius. If the given temperature has two significant figures, the answer is 9000°F. The addition or subtraction of 32 can often be omitted in conversions involving high temperatures.

(b) 104°F   (c) 414°F   (d) −40°F.

B-6 (a) 5273 K   (b) 313.2 K   (c) 485 K   (d) 233 K.

B-8 (a) 6.82 MPa   (b) 1.0×10^6 V m^{-1} (c) 2.3×10^{-17}m s^{-1} (d) 2.24×10^{-2} m^3 mol^{-1}
(e) The answer is 1.03×10^4 kg m^{-2}. Part (e) may be harder for those with experience than for beginners. If "14.7 lb in^{-2}" is recognized as ordinary atmospheric pressure, then the answer 1.01×10^5 Pa may be written immediately. But the use of the pound as a unit of force has been deliberately avoided in the Appendix.

B-10 The unit-factor method gives:

$$1\,\frac{\text{mile}}{\text{gal}} \times \left(\frac{1\ \text{gal}}{3.785\ \text{dm}^3}\right) \times \left(\frac{1609.344\ \text{m}}{1\ \text{mile}}\right) = 425.2\frac{\text{m}}{\text{dm}^3}$$

Therefore, 30.0 miles per gallon is 1.28×10^4 m dm^{-3}. Fuel consumption could also be given in the sleek but opaque unit "m^{-2}".

B-12 (a) The conversion factor between miles and meters is in problem B-10. The answer is 1.86×10^5 miles per second.

(b)

$$3.00 \times 10^8 \frac{\text{m}}{\text{s}} \times \frac{86400\ \text{s}}{1\ \text{day}} \times \frac{14\ \text{days}}{1\ \text{fortnight}} \times \frac{1\ \text{in}}{0.0254\ \text{m}} \times \frac{1\ \text{ft}}{12\ \text{in}} \times \frac{1\ \text{furlong}}{660\ \text{ft}}$$

$$= 1.80 \times 10^{12} \text{ furlongs per fortnight}$$

Appendix C

C-2 The slope is 0.0146 atm $(°C)^{-1}$.

C-4 (a) The equation is in the required form with m the slope equal to -2 and b the intercept equal to -8.

(b) The equation is $y = 3/4x + 7/4$. The slope is 3/4 and the intercept is 7/4.

(c) The equation is $y = 16/7x - 53/7$. The slope is 16/7 and the intercept is $-53/7$.

C-6 The graph of y versus x for the equation

$$y = \frac{8 - 10x - 3x^2}{2 - 3x}$$

is linear. Note that the numerator can be factored into $(2 - 3x)(4 + x)$ so that the equation is equivalent to $y = 4 + x$.

C-8 (a) $x = 3/4$ (b) $x = -5/2$ (c) $x = -2$.

C-10 The answers are given to 5 significant figures. (a) $x = -0.14132, -2.3587$ (b) $x = -0.47178, 1.2718$ (c) $x = 1.0000, 2.3333$.

C-12 (a) Assuming that x is small compared to 2.00 gives $x = 5.27 \times 10^{-17}$. The other roots are $x = -2.00, x = -2.50, x = 3.00$, as can be confirmed by approximating the right side of the equation as zero.

(b) The best method of solution is graphical. There are three roots because this is a third-degree equation: $x = 0.0827, 0.967, -3.05$.

(c) The only real root is $x = 0.287$. It can be arrived at graphically. The other two roots are imaginary: $x = 0.023155 \pm 1.7028i$. They are of little interest in chemical applications.

C-14 (a) 2.96×10^{-17} (b) 31.392   (c) 1.5×10^6 (d) -10.3025

C-16 1.342×10^{-7}.

C-18 Few calculators accommodate a number with an exponent exceeding 99 in absolute value. To answer this problem, write $10^{-107.8} = 10^{-108} \times 10^{+.2}$. Then, compute the parts of the product separately. The answer is 2×10^{-108}.

C-20 The two logarithms are 192.428 and -288.572.

C-22 The problem is to find x in the equation $x = 1/\ln x$. One way to proceed is to guess an x, put it into the right side of the equation and see on a calculator if the indicated operations gives back the guess. The answer is 1.763.

C-24 (a) $\dfrac{dy}{dx} = 114x^{18}$

(b) $\dfrac{dy}{dx} = 14x + 6$

(c) $\dfrac{dy}{dx} = -6e^{-6x}$

(d) $\dfrac{dy}{dx} = -2\sin 2x - \dfrac{7}{x^2}$

C-26 (a) $\displaystyle\int_{-1}^{3} 4\,dx = 4x]_{-1}^{3} = 12 - (-4) = 16$

(b) $\displaystyle\int_{2}^{100} \dfrac{1}{x}\,dx = \ln x]_{2}^{100} = \ln 100 - \ln 2 = \ln 50 = 3.912$

(c) $\displaystyle\int_{2}^{4} \dfrac{5}{x^2}\,dx = -\dfrac{5}{x}]_{2}^{4} = -\dfrac{5}{4} - \left(-\dfrac{5}{2}\right) = 1.25$